Field Manual of Diseases on Fruits and Vegetables

R. Kenneth Horst

Field Manual of Diseases on Fruits and Vegetables

Springer

R. Kenneth Horst
Plant Pathology and
Plant Microbe Biology
Cornell University
Ithaca, NY, USA

ISBN 978-94-007-9385-9 ISBN 978-94-007-5974-9 (eBook)
DOI 10.1007/978-94-007-5974-9
Springer Dordrecht Heidelberg New York London

Softcover re-print of the Hardcover 1st edition 2013

Printed on acid-free paper

Springer is part of Springer Science+Business Media (www.springer.com)

Preface

The information in this manual is taken in large part from the records of the Plant Disease Survey as given in the *Plant Disease Reporter*, *Plant Disease*, and from the *Index of Plant Diseases in the United States*, *Agriculture Handbook 165*, U.S. Department of Agriculture. There are a great many records that bring in new reports every day, and any such list is out of date by the time it is published, and is sure to be far from complete. This check list is offered as a helpful guide, a foundation on which to build. It is by no means the last word.

Hosts are listed alphabetically by common names except where the scientific names mean less confusion. The diseases are those mostly reported from North, Central, and South America.

Contents

Part I

Illustrations

Plate 1 Late Blight on Potato

R.K. Horst, *Field Manual of Diseases on Fruits and Vegetables*,
DOI 10.1007/978-94-007-5974-9_1,

Plate 2 Late Blight on Tomato

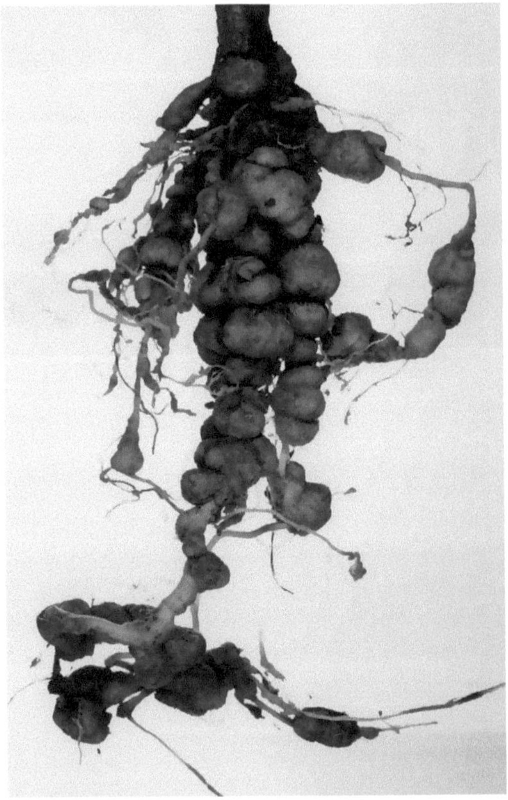

Plate 3 Root-Knot Nematode; pear-shaped female with egg sac; encysted young larva; and worm-like adult male

Plate 4 Corn Smut

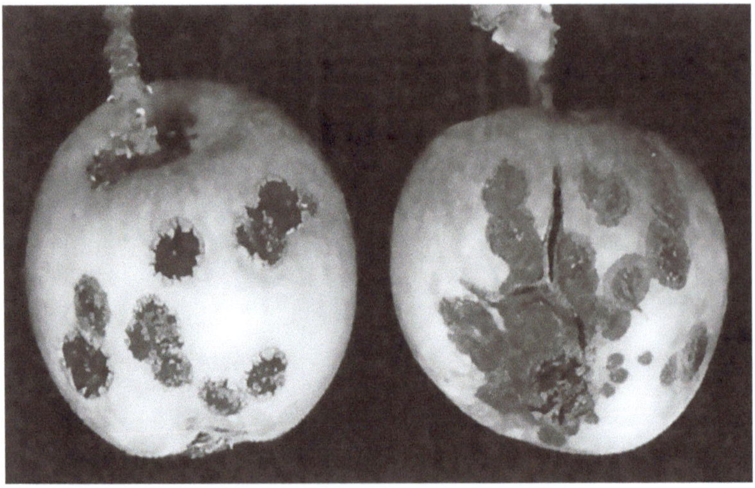

Plate 5 Apple Scab on Leaf and Fruit

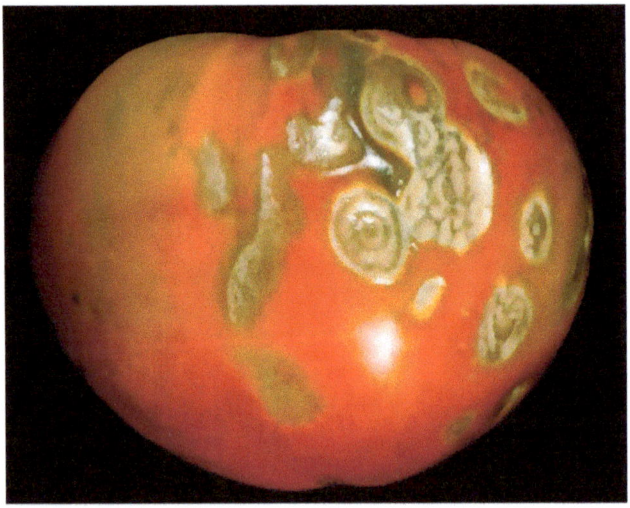

Plate 6 Ringspot on Tomato Fruit caused by Cucumber Mosaic Virus

Plate 7 Late Blight on Potato caused by **Phytophthora infestans**

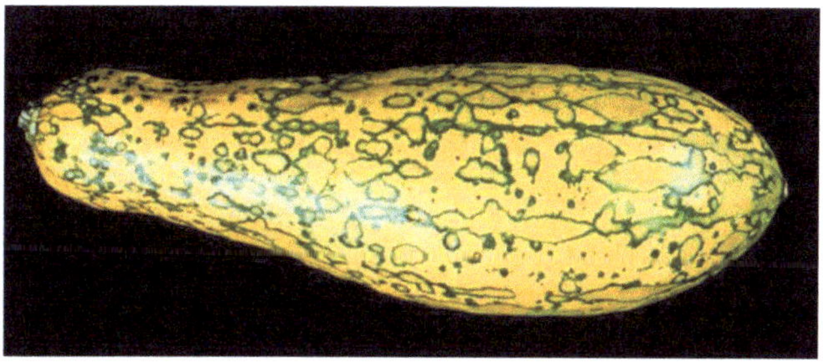

Plate 8 Mosaic on Squash caused by Cucumber Mosaic Virus

Plate 9 Apple Scab on Apple caused by **Venturia inaequalis**

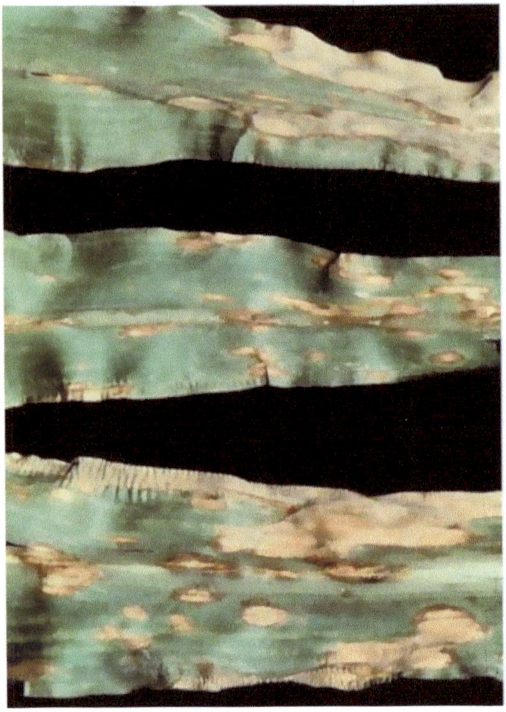

Plate 10 Yellow Leaf Blight on Corn caused by **Phyllosticta maydis**

Plate 11 Scab on Citrus caused by **Elsinoë fawcettii**

Plate 12 Brown Rot on Cherry caused by **Monilinia fructicola**

Part II

Disease Control and Management

Control of a plant disease means reduction in the amount of damage caused. Our present annual toll from disease is nearly four billion dollars. Perfect control is rare, but profitable control, when the increased yield more than covers the cost of chemicals and labor, is quite possible. Commercial growers now average a return of four dollars for each dollar so invested. Keeping home plantings ornamental yields a large return in satisfaction and increased property value.

The five fundamental principles of control are exclusion, eradication, protection, resistance, and therapy.

1. *Exclusion* means preventing the entrance and establishment of pathogens in uninfested gardens, states, or countries. For home gardeners it means using certified seed or plants, sorting bulbs before planting, discarding any that are doubtful, possibly treating seeds or tubers or corms before they are planted, and, most especially, refusing obviously diseased specimens from nurseryman or dealer. For states and countries, exclusion means quarantines, prohibition by law. Sometimes restricted entry of nursery stock is allowed, the plants to be grown in isolation and inspected for 1 or 2 years before distribution is permitted.
2. *Eradication* means the elimination of a pathogen once it has become established on a plant or in a garden. It can be accomplished by *removal* of diseased specimens, or parts, as in roguing to control virus diseases or cutting off cankered tree limbs; by *cultivating* to keep down weed hosts and deep ploughing or spading to bury diseased plant debris; by *rotation* of susceptible with nonsusceptible crops to starve out the pathogen; and by *disinfection*, usually by chemicals, sometimes by heat treatment. Spraying or dusting foliage with sulfur after mildew mycelium is present is eradication, and so is treating the soil with chloropicrin to kill nematodes and fungi.
3. *Protection* is the interposition of some protective barrier between the susceptible part of the suscept or host and the pathogen. In most instances this is a protective spray or dust applied to the plant in advance of the arrival of the fungus spore; sometimes it means killing insects or other inoculating agents; sometimes it means the erection of a windbreak or other mechanical barrier.

R.K. Horst, *Field Manual of Diseases on Fruits and Vegetables*,

DOI 10.1007/978-94-007-5974-9_2, © Springer Science+Business Media Dordrecht 2013

4. *Resistance* is control by the development of resistant varieties. Resistant varieties are as old as time. Nature has always eliminated the unfit, but since about 1,890 man has been speeding up the process by deliberately breeding, selecting, and propagating plants resistant to the more important diseases. Resistant ornamental plants have lagged behind food plants, but we do have wilt-resistant asters, rust-resistant snapdragons, wilt-resistant mimosas. Here is the ideal way for home gardeners to control their plant diseases – in the winter when the seed order and the nursery list is made out – so easy, and so safe!
5. *Therapy* is control by inoculating or treating the plant with something that will inactivate the pathogen. Chemotherapy is the use of chemicals to inactivate the pathogen, whereas heat is sometimes used to inactivate or inhibit virus development in infected plant tissues so that newly developing tissue may be obtained which is free of the pathogen.

A **fungicide** is a substance that destroys or inhibits the growth of fungi. It may be an **eradicant**, applied to a plant, plant part, or the environment as a curative treatment to destroy fungi established within a given area or plant; or preferably it may be a **protectant**, applied to protect a plant or plant part from infection by killing, or inhibiting the development of, fungal spores of mycelium that may arrive at the infection court. A **bactericide** is a substance that destroys or inhibits bacteria and nematicide for nematodes. Among the more recent bactericides are antibiotics, products of other living organisms. They also have value against certain fungi. There are few **virocides**, which are toxic or poisonous to viruses.

A **pesticide** is any chemical that is used to kill pests, especially insects and rodents.

An **insecticide** is an agent used to kill insects.

A **disinfectant** is an agent that frees a plant or plant part from infection by destroying the pathogen established within it. A disinfectant kills or inactivates organisms present on the surface of the plant or plant part or in the immediate environment. Chemicals for seed treatment can be either eradicants or protectants, but most of them are disinfestants, in that they kill organisms on the surface of the seed rather than those within. In common usage, however, they are called disinfectants.

A **nematicide** is, of course, a chemical that kills nematodes in the soil or in the plant. Most nematicides are **fumigants**, chemical toxicants that act in volatile form.

The search for new fungicides goes on, with hundreds of synthetic organic compounds being screened each year. This screening is often a cooperative venture between manufacturers, state experiment stations, and the U.S. Department of Agriculture. After safety precautions for the operator and the environment, and the effectiveness of a compound for certain diseases have been determined, the chief question is whether the material is **phytotoxic**, that is, injurious to plants, at concentrations required for control. Phytotoxicity is an elusive factor, not to be pinned down in a few tests. It varies not only with the kind of plant but with the particular variety, the amount of moisture in the soil when the spray is applied, the temperature, whether or not the application is followed by rain or high humidity, the section of the country, and the compatibility of the chemical with spreaders or wetting agents, as well as with other fungicides or insecticides. Coordinated tests with new materials in many different states are extremely valuable. Some compounds give rather uniform results over the country; others vary widely with climatic conditions.

Consumers are well protected against fraud, but they must be willing to read the fine print on labels if they are to choose intelligently from the bewildering array of proprietary compounds on dealers' shelves. Consumers must also read the fine print and follow directions exactly if their homegrown vegetables are to be as safe for consumption as those from commercial growers who have to comply with the law in the matter of residues.

Even if you follow exactly the directions for dosage given on the label, you may have some plant injury under your particular combination of soil, weather, and kinds of plants. Keep a notebook. Put down the date you sprayed, the dosage used, the approximate temperature and humidity, whether it was cloudy or sunny, in a period of drought or prolonged wet weather. Go around later and check for burning; for leaf spotting and defoliation from the spray or from failure to control the disease; for leaf curling or stunting; for too much unsightly residue. Note which varieties can take the spray and which cannot. Exclusive reliance must be placed on directions and information supplied by the manufacturer or by agricultural specialists, agents, or advisors. **Be sure to read the label**.

Part III

Host Plants and Their Diseases

Headings Under Which Diseases Are Described

Anthracnose – dead spots with definite margins, often with pinkish slimy spore masses, on leaves, stems, or fruit.

Bacterial Diseases – all types of diseases, galls, blights, rots, leaf spots, caused by bacteria.

Black Knot – black, knotty enlargement of woody tissue.

Blackleg – darkening at the base of a plant.

Black Mildew – superficial dark growth caused by parasitic fungi.

Blackspot – a dark leaf spot on rose.

Blights – general killing of leaves, flowers, stems.

Blotch Diseases – irregular necrotic areas on leaves or fruit.

Broomrapes – leafless herbs parasitic on roots.

Cankers And Dieback – localized lesions on stems or trunks, sometimes accompanied by dying back from the top.

Club Root – distorted swollen roots.

Damping-Off – sudden wilting of seedlings or rotting of seeds in soil.

Dodder – parasitic seed plant with orange tendrils.

Downy Mildews – with internal mycelium but fruiting structures protruding to form white, gray, or violet patches.

Fairy Rings – mushrooms growing in circles.

Fruit Spots – blemishes on fruit.

Galls – noticeable enlargements of leaves, stems, or roots.

Leaf Blister, Leaf Curl Diseases – leaf deformities.

Leaf Scorch – discoloration as if by intense heat.

Leaf Spots – delimited dead areas in leaves.

Lichens – occasional on trees, or shrubs.

Mistletoe – semiparasitic seed plant, forming leafy tufts in trees.

Molds – conspicuous fungus growth on leaves, seeds, or grafts.

Needle Casts – conspicuous shedding of evergreen foliage.

Nematodes – causing decline diseases and flowers.

R.K. Horst, *Field Manual of Diseases on Fruits and Vegetables*,
DOI 10.1007/978-94-007-5974-9_3, © Springer Science+Business Media Dordrecht 2013

NONPARASITIC DISEASES – due to environmental conditions rather than specific organisms.

POWDERY MILDEWS – superficial white felty or powdery growth on leaves and flowers.

ROTS – soft or hard decay or disintegration of plant tissues.

RUSTS – with reddish or rust-colored spore masses.

SCAB – raised or crustlike lesions on leaves or fruit.

SCURF – flaky or scaly lesions.

SLIME MOLDS – found in lawns.

SMUTS – with sooty black spore masses.

SNOW MOLD – light patches in turf, especially early spring.

SOOTY MOLD – superficial black mycelium growing in insect exudate.

SPOT ANTHRACNOSE – light spots with raised darker borders or scabby lesion caused by *Elsinoë* species.

VIRUS DISEASES – mosaics, ring spots, yellows, wilt caused by viruses.

WHITE RUSTS – white blisters in leaves.

WILTS – systemic diseases, with wilting, death of leaves, and branches.

WITCHWEED – weed parasitic on roots. The numbers given with the rusts refer to spore stages.

A

ALMOND (*Prunus amygdalus*)

ANTHRACNOSE. **Colletotrichum acutatum**, CA.

ANTHRACNOSE; Kernel Rot. **Gloeosporium amygdalinum**, CA.

BACTERIAL Crown Gall. **Agrobacterium tumefaciens**, AL, AZ, CA, NC.

BACTERIAL Leaf Spot. **Xanthomonas pruni**, NE.

BACTERIAL Shoot Blight; Blast. **Pseudomonas syringae**, CA.

BLIGHT, Blossom; Brown Rot. **Monilinia laxa**, CA, OR; **Botrytis cinerea**.

BLIGHT, Leaf; Shot-Hole Disease. **Coryneum carpophilum**, CA, ID, OR.

CANKER; DIEBACK; Crown Rot. **Phytophthora cactorum**, CA, **P. citrophthora**, CA; **Botryosphaeria dothidea** (Dothiorella canker) CA; **Ceratocystis fimbriata**, CA.

CANKER, Pruning Wound. **Phytophthora syringae**, CA.

DECLINE, Almond. Unknown etiology, CA. "Golden Death".

LEAF SPOT. **Cercospora circumscissa**, CA, OR; **Hendersonia rubi**, CA; **Alternaria alternata**, CA; **A. arborescens**, CA; **A. tenuissima**, CA.

NEMATODE, Lesion. **Pratylenchus vulnus**, CA.

NEMATODE, Root Knot. **Meloidogyne** sp., AZ.

NONPARASITIC. **Bud Failure**. Seed transmitted, increased by pruning.
Chlorosis. Iron deficiency.
Little Leaf. Zinc deficiency.

POWDERY MILDEW. **Podosphaera tridactyla**, CA; **Sphaerotheca pannosa**, CA.

ROT, Fruit. **Phomopsis amygdali**, CA, and associated branch dieback.

ROT, Green Fruit. **Sclerotinia sclerotiorum**, CA.

ROT, Heart. **Polyporus versicolor**, OR.

ROT, Hull. **Rhizopus** spp.

ROT, Root. **Armillaria mellea**, CA, NC; **Phymatotrichum omnivorum**, TX.

ROT, Wound. **Schizophyllum commune**, CA.

RUST. **Tranzschelia discolor** (II, III); 0, I on anemone, CA.

SCAB. **Cladosporium carpophilum**, CA, CT, OR.

R.K. Horst, *Field Manual of Diseases on Fruits and Vegetables*,
DOI 10.1007/978-94-007-5974-9_4,

VIRUS. **Almond Bud Failure**; **Almond Calico**, CA; **Peach Mosaic**; **Peach Ring Spot**; **Peach Rosette**; **Peach Yellow Bud Mosaic**; **Peach Yellows**.
WILT. **Verticillium albo-atrum**, CA, IL.

Crown rot and Armillaria rot are often limiting factors in almond production. At least five sprays are necessary to control brown rot, leaf blights, and scab, and often a zinc sulfate spray for little leaf. The California Agricultural Experiment Station Extension Service provides, each year, a revised "Spray, Dust and Fumigation Program for Almonds".

ANTHRISCUS (Chervil)
POWDERY MILDEW. **Erysiphe heraclei**, CA.
ROT, Stem and Crown. **Sclerotinia sclerotiorum**, CA.

APPLE (*Malus sylvestris*)
ANTHRACNOSE, Northwestern; canker; Fruit Rot. **Neofabraea malicorticis**, prevalent OR, WA, occasional CA, IL, MA, ME, MI, NE, OK.
BACTERIAL Blast, of flowers, shoots. **Pseudomonas syringae**, AR, CA; Blister Spot of bark and fruit. **P. syringae** pv. **papulans**; Rot, **P. melophthora**.
BACTERIAL Crown Gall. **Agrobacterium tumefaciens**, general; Hairy Root, **A. rhizogenes**, Central States, ID, NY.
BACTERIAL Fire Blight. **Erwinia amylovora**, general.
BLIGHT, Limb. **Corticium laetum**, LA, NC; **C. salmonicolor**, FL, LA.
BLIGHT, Southern. **Sclerotium rolfsii**, on seedlings.
BLIGHT, Thread. **Pellicularia koleroga**, WV and IN to Gulf states.
BLOTCH, Fruit; LEAF SPOT; Twig Canker. **Phyllosticta solitaria**, general except New England and the far South.
BLOTCH, Sooty, of fruit. **Gloeodes pomigena**, eastern and central states; **Geastrumia polystigmatis**, NC.
CANKER, Bark; Fruit Rot. **Myxosporium corticola**, New England to MD, IL, MI, OK, OR, SD.
CANKER, Blister. **Nummularia discreta**, east of the Rocky Mts.
CANKER, Bole. **Cylindrocarpon didymium**, WA; **Leucostoma cincta**. WI.
CANKER, Crater. Cause unknown, sunken bark, crown rot, decline. WA.
CANKER; DIEBACK. **Botryosphaeria ribis** (syn. **B. dothidea**), East and South; **Cytospora** spp., widespread; **Fusarium** spp., also Fruit Rot, Pacific Northwest, Secondary to drought, winter injury, insect punctures; **Coryneum foliicola**, widespread; **Glutinium macrosporium**, OR; **Leptosphaeria coniothyrium**, East, central states, WI; **Plenodomus fuscomaculans**, CA, MI; **Sphaeropsis pyriputrescens**, WA.
CANKER, European. **Nectria galligena**, East and central states to NC and MS; Pacific Coast.
CANKER; Leaf Spot. **Monochaetia mali**, IL, NJ to SC, MO, WV; **Leucostoma cincta**, MI.
CANKER, Perennial; Bull's-Eye Fruit Rot. **Neofabraea perennans**, ID, MT, OR, WA.
CANKER; Sapwood Rot. **Hymenochaete agglutinans**, CT; **Hypoxylon** spp.

CANKER, Scurf Bark. **Clasterosporium** sp.

CANKER, Twig. **Coryneum foliicola**, widespread; **Gibberella baccata**; **Nectria cinnabarina**, northern U.S.

DECLINE. **Irpex tulipiferae**, MN; **Coriolus versicolor**, also dieback, MN, WA; **Schizophyllum commune**, MN.

FRUIT SPOT, Black Pox; Canker; Leaf Spot. **Helminthosporium papulosum**, GA, IN, MA, MS, NJ, OH, PA, WV; **Phacidiopycnis washingtonensis** (also Fruit Rot), WA.

FRUIT SPOT, Brooks. **Mycosphaerella pomi**, New England to NC, AR, IA, MO.

FRUIT SPOT, Flyspeck. **Microthyriella rubi**, general.

LEAF SPOT. **Cercospora mali**, Gulf states, VT; **Coniothyrium pyrinum**; **Diaporthe pernisiosa**, AR; **Fabraea maculata**, IA; **Illosporium malifoliorum**, PA to NC, IN; **Mycosphaerella sentina**, IL, NJ, PA; **Pestalotia** spp., MD to NC, ID, IN; Ghost Spot, **Alternaria tenuis**, GA; **Cristulariella pyramidalis**, FL, on *Malus pumila*, **Glomerella cingulata**, TN.

MISTLETOE. **Phoradendron serotinum (flavescens)**, NC, TX; **Viscumalbum**, CA.

NEMATODE, Lesion. **Pratylenchus penetrans**; **P. pratensis**.

NEMATODE, Root Knot. **Meloidogyne** spp., MS, TX, UT.

NONPARASITIC. **Bitter Pit**, **Stippen**, **Baldwin Spot,** general in storage.

Black End. Probably drought injury, AL, OR, WA.

Black Heart, of wood. Perhaps freezing injury; widespread.

Brown Core, in MacIntosh variety. Excessive nitrogen, low storage temperature.

Callus Knot. Wound overgrowth from defective union of stock and scion.

Chlorosis. Usually iron deficiency in alkaline western soils.

Collar Rot. Winter injury.

Cork; **Rosette**. Boron deficiency, northeastern states to IN, KY, Pacific Coast.

Internal Bark Necrosis. Cause unknown, possibly same as measles; general.

Jonathan Spot. Associated with dry weather, delayed storage.

Leaf Scorch. Magnesium deficiency.

Measles, reddish pimples in bark. Widespread; cause unknown, possibly boron deficiency in part. False measles due to manganese toxicity.

Ozone Injury. Pitted area in flesh.

Rosette; Little Leaf. Usually zinc deficiency.

Scald. Discoloration of fruit skin by volatile respiratory products, general in storage. Controlled by oiled paper wraps.

Soft Scald and **Soggy Breakdown**. Associated with delayed storage and low temperature.

Stigmonose. Fruit dimpling, distortion from insect punctures; widespread.

Sunburn, of fruit. Heat or light injury.

Sunscald, of bark. Freezing injury of trunk and larger branches on side exposed to sun.

Water-Core. Deficient or irregular water supply.

POWDERY MILDEW. **Podosphaera leucotricha** and **P. oxyacanthae**, general.

ROT, Bitter, of Fruit; Canker. **Glomerella cingulata**, general, especially in South, rare in West; **G. rubicola**, IL; **Colletotrichum acutatum**, AR, MI.

Rot, Black; Frog-Eye Leaf Spot; Canker. **Physalospora obtusa**, general to the Great Plains; **P. mutila**, CA, MT, OR, WA; **P. rhodina**, KY.

Rot, Black Root. **Botryosphaeria obtusa**, PA; **Xylaria mali**, eastern and central states.

Rot, Brown. **Monilinia fructicola**, general except for far South; **M. laxa**, OR, WA.

Rot, Calyx-end. **Alternaria** sp., **Sclerotinia sclerotiorum**, NH, WA.

Rot, Crown. **Phytophthora cactorum**, MA.

Rot, Fruit. **Alternaria mali**; **Aspergillus niger**; **Cephalosporium carpogenum**; **Corticium centrifugum**, Pacific Northwest, occasional in East; **Chaetomella** sp., WA; **Cladosporium** spp.; **Endomyces mali**; **Epicoccum granulatum**, WA; **Gloeosporium** spp.; **Gliocladium viride**, IL; **Hormodendron cladosporiodes**, WA; **Mucor piriformis**, CA, WA; **Oospora** sp.; **Phialophora malorum**, IN, VA, WA; **Pleospora fructicola**, WA; **P. herbarum**, Pacific states; **Penicillium** spp., cosmopolitan; **Phoma** spp.; **Stemphylium congestum**, WA; **Tricothecium roseum**, general; **Botrytis** spp.; **Colletotrichum fructus**; **Rhizopus nigricans**, cosmopolitan; **Phytophthora syringae**, OR.

Rot, Fruit (Postharvest). **Botrytis cinerea**, **Phomopsis mali**, core rot, CA, NY; **Trichoderma harzianum**, MD; **Sphaeropsis pyriputrescens**, WA; **Phacidiopycnis washingtonensis,** WA; **Potebniamyces pyri**, WA.

Rot, Heartwood. **Fomes fomentarius**; **F. igniarius**; **F. pinicola**; **Ganoderma applanatum**.

Rot, Mushroom Root. **Armillaria mellea**, prevalent on Pacific Coast; **Clitocybe tabescens**, AR, FL, TX, VA.

Rot, Postharvest Speck. **Phacidiopycnis washingtonensis**, WA.

Rot, Root. **Botryodiplodia theobromae**, AL; **Hymatotrichum omnivorum**, AK, AZ, NM, TX; **Fusarium oxysporum**, ID; **F. solani**, ID.

Rot, Silverleaf. **Stereum purpureum**, KS, ME, MN, NY, Pacific Northwest.

Rot, White Root. **Corticium galactinum**, DE to VA; AR, IL; **Rosellinia necatrix**, CA.

Rot, Wood. **Daedalea confragosa**; **Poria** spp.; **Pleurotus ulmarius**; **Pholiota adiposa**; **Polyporus** spp.; **Stereum** sp.; **Schizophyllum commune**; **Trametes** spp.

Russet, Fruit. **Aureobasidium pullulans** and **Rhodotorula glutinis**, NY.

Rust. **Gymnosporangium libocedri**. (0, I), on leaves, fruit; III on incense cedar, CA, OR; **Gymnosporangium nidus-avis** (0, I), on leaves, fruit, stems, IN, MD, MS, NJ; III on red-cedar.

Rust, Cedar-Apple. **Gymnosporangium juniperivirginianae** (0, I) on leaves, fruit; general east of Great Plains; III on red-cedar and Rocky Mountain juniper.

Rust, Hawthorn. **Gymnosporangium globosum** (0, I) on leaves; III on red-cedar, ME to AK, KS, NE.

Rust, Quince. **Gymnosporangium clavipes** (0, I), on fruit; III on common juniper and red-cedar.

Scab. **Venturia inaequalis**, general.

Sooty Mold. **Fumago vagans**, occasional.

Spot Anthracnose. **Elsinoë piri**, OR, WA.

VIRUS. **Apple Chlorotic Leaf Spot**; **Apple Stem Growing**; **Apple Mosaic**; **Tulare Apple Mosaic**, CA; **Tobacco Mosaic**; **Dapple Apple**, NH; **Green Mottle**, NY; **Stem-pitting**; **Tomato Ringspot,** NY; **Decline** and **Graft Union Necrosis**, WA.
WILT. **Verticillium albo-atrum**.

If this appalling list of diseases should make you think twice before planting apples in the backyard with the expectation of getting cheap and abundant fruit, that is all to the good. There is no easy, or cheap, road to perfect fruit. The commercial grower may, in a wet season, apply nearly 20 sprays to keep scab under control. The homeowner thinks one or two are enough. State experiment stations offer abbreviated schedules for home gardeners, tailored for the area, and if these are followed carefully, a fairly good crop can be expected. There are also all-purpose mixtures available for fruit trees, which may work reasonably well. Scab is the most important apple disease, and proper timing of early season sprays is most essential. Bitter rot, black rot, sometimes rust need attention. Fire blight control is primarily a question of proper pruning to remove infected wood, with an antibiotic spray during bloom.

APRICOT (*Prunus armeniaca*)
BACTERIAL Canker; Gummosis. **Pseudomonas syringae**, CA, OR; **Criconema mutabile**, CA.
BACTERIAL Crown Gall. **Agrobacterium tumefaciens**, widespread. Japanese apricot (*Prunus mume*) is resistant.
BACTERIAL Fire Blight. **Erwinia amylovora**, CO, FL, NE, PA, TX.
BACTERIAL Leaf Spot **Xanthomonas pruni**, IL to TX, NE.
BLACK KNOT. **Dibotryon morbosum**, CO, IA, NY.
BLIGHT, Blossom, Twig; Brown Rot. **Monilinia laxa**, Pacific Coast states.
BLIGHT, Shoot; Shot Hole; Fruit Spot. **Coryneum carpophilum**, wide-spread; **Monilinia laxa**, MA, NY, RI.
CANKER, Felt Fungus. **Septobasidium pseudopedicellatum**, MS.
CANKER, Trunk. **Phytophthora cactorum**, CA; **P. citrophthora**, CA.
CANKER, Trunk and Limb Gall. **Monochaetia rosenwaldia**, CA.
CANKER, Twig; DIEBACK. **Cytospora** spp., AZ, NY; **Valsa leucostoma**, MO, TX; Cytosporina Dieback, **Eutypa armeniaceae** (syn. **E. lata**), CA; **Ceratocystis fimbriata**, CA.
CANKER; DIEBACK; Coral Spot. **Nectria cinnabarina**, IN, WA.
FRUIT SPOT. **Venturia cerasi** (See under SCAB).
LEAF CURL. **Taphrina deformans**, SC.
LEAF SPOT. **Cercospora circumscissa**, TX; **Coccomyces** sp., CA, MA, TX, VT; **Phyllosticta circumscissa**, CA.
NEMATODE, Dagger. **Xiphinema** sp., WV, NY.
NEMATODE, Lesion. **Pratylenchus vulnus**, CA; **P. penetrans**, NY.
NEMATODE, Root Knot. **Meloidogyne** sp., AZ, TX.
NONPARASITIC. **Arsenical Injury**, from accumulation in soil, CA.
Chlorosis, alkali injury. Mineral deficiency, AZ, CA.
Exanthema. Copper deficiency, CA.
Gummosis, Sour Sap. Adverse soil and moisture conditions, AZ, CA, NJ, WA.

Internal Browning, Fruit Cracking. Boron deficiency, WA.
Little Leaf. Zinc deficiency, CA, OR, WA.
Mottle Leaf. Manganese deficiency, CA.

POWDERY MILDEW. **Podosphaera oxyacanthae**, CA, IA; **Sphaerotheca pannosa**, CA, NY.
ROT, Blossom End; Fruit Spot. **Alternaria** sp., and **A. citri** CA.
ROT, Brown; Blossom Blight. **Monilinia fructicola**, widespread on ripe fruit.
ROT, Green Fruit; Gummosis. **Botrytis cinerea**, CA.
ROT, Green Fruit; Twig Blight. **Sclerotinia sclerotiorum**, CA; **Lambertella pruni**, CA.
ROT, Heart. **Schizophyllum commune**, after freezing, TX, WA; **Trametes hispida**, CO.
ROT, Mushroom Root. **Armillaria mellea**, CA, TX; **Clitocybe tabescens**, FL.
ROT, Root. **Phymatotrichum omnivorum**, AZ, OK, TX.
ROT, Silver Leaf. **Stereum purpureum**, CA.
RUST. **Tranzschelia discolor** and **T. pruni** (II, III), CA, MS, NM, TX.
SCAB; Freckle; Twig Canker. **Cladosporium carpophilum**, wide spread.
VIRUS. **Apricot Gummosis**, WA; **Apricot Pucker Leaf**, UT; **Apricot Ring Pox**, CA, CO, WA; **Peach Mosaic**, Southwest; **Peach Phony**, GA; **Peach Ring Spot**, western U.S.; **Peach Rosette**; **Peach Yellow Bud Mosaic**, CA; **Peach Yellows**, occasional in East; Stem Pitting.
WILT. **Verticillium albo-atrum**, CA, UT, WA.

Apricots are very susceptible to Armillaria root rot and should be on resistant Myrobalan rootstock. Bacterial canker is an epidemic disease in many seasons, with activity starting in late autumn, ceasing in early summer. Sprays are usually needed in California for zinc deficiency, brown rot, green or jacket rot, and shot hole.

ARTICHOKE, GLOBE (Cynara scolymus)
BLIGHT, Gray Mold. **Botrytis cinerea**, CA, NY.
BLIGHT, Southern. **Sclerotium rolfsii**, GA.
LEAF SPOT. **Cercospora obscura**, CA, TX; **Cladosporium** sp., CA, SC; **Ramularia cynarae**, CA, prevalent.
NEMATODE, Root Knot. **Meloidogyne** sp., CA.
POWDERY MILDEW. **Golovinomyces cichoracearum**, CA, NJ.
ROT, Root. **Phytophthora megasperma**, CA; **Phymatotrichum omnivorum**, TX.; **Pythium aphanidermatum**, CA.
Rot, Root and Stem. **Rhizoctonia solani**, MS, TX; **Sclerotinia sclerotiorum**, OR.
Virus. **Artichoke Curly Dwarf**; Artichoke Latent, CA.
WILT. **Verticillium dahliae**, CA.

ASPARAGUS (Asparagus officinalis)
ANTHRACNOSE; Canker. **Colletotrichum** sp., AL, CT, IL.
BACTERIAL Soft Rot. **Erwinia carotovora**, general.
BLIGHT, Ashy Stem. **Macrophomina phaseoli**, TX.
BLIGHT, Branchlet; DIEBACK. **Alternaria** sp., IL, MA, NY, OK, SC; **Stemphylium botryosum**, secondary; Ascochyta sp., DE, TX.
BLIGHT, Gray Mold, Shoot. **Botrytis cinerea**, CA, IL, MA, NY, WV.

CANKER; Stem Spot. **Phoma asparagi**.
DAMPING-OFF; Stem Canker. **Rhizoctonia solani**, occasional.
LEAF SPOT. **Cercospora asparagi**, general; **Pleospora herbarum**; **P. allii** Anamorph, **Stemphylium vesicarium**, MI, OK, WA.
NEMATODE, Root Knot. **Meloidogyne** sp., SC. Usually resistant.
ROT, Crown. **Penicillium martensii**, WA; **Penicillium** sp., blue mold rot.
ROT, Mushroom Root. **Armillaria mellea**, OR.
ROT, Root; Stem Wilt; Decline. **Fusarium oxysporum** f. sp. **asparagi**, CA, SC, WA; **Fusarium proliferatum**, CT; **Fusarium** sp.; **Phytophthora** sp., MI; **P. megasperma**, Montreal, Canada; P. **asparagi**, MI.
ROT, Stem. **Diplodia asparagi**; **Phytophthora** sp., CA.
ROT, Watery Soft. **Sclerotinia sclerotiorum**, occasional in South.
RUST. **Puccinia asparagi** (0, I, II, III), general on susceptible varieties. Rust is the most important asparagus disease, and resistant varieties are sometimes disappointing.
VIRUS. Asparagus I, DE, MI, NJ; Asparagus II, DE; Tobacco Streak.

AVOCADO (Persea americana)
ANTHRACNOSE; Leaf and Fruit Spot; Black Rot. **Glomerella cingulata**, Mexico, general; **Colletotrichum gloeosporioides**, Mexico.
BACTERIAL CANKER. **Xanthomonas campestris**, CA.
BACTERIAL Crown Gall. **Agrobacterium tumefaciens**, CA.
BACTERIAL Fruit Blast. **Pseudomonas syringae**, CA.
BLACK MILDEW. **Irene perseae**, FL.
BLACK STREAK. No causal agent has been identified, FL.
BLIGHT, Seedling. **Sclerotium rolfsii**, FL; **Phytophthora palmivora**, FL.
BLOTCH, Fruit Spot. **Cercospora perseae**, FL; **C. purpurea**.
CANKER, Branch; Dothiorella Fruit Rot. Botryosphaeria ribis var. chromogena, CA; **Neofusicoccum austral, N. luteum, and N. parvum**, CA.
LEAF SPOT. **Phyllosticta micropuncta**, AL, FL.
LEAF SPOT; Algal; Green Scurf. **Cephaleuros virescens**, FL.
LEAF SPOT; Fruit Spot; Seedling Blight. **Pestalotia** spp., general.
LEAF SPOT; Smudgy Spot on Twigs. **Helminthosporium** sp., CA.
NEMATODE, Burrowing. **Radopholus similis**, FL.
NEMATODE, Dagger. **Xiphinema americanum**.
NEMATODE, Meadow. **Pratylenchus brachyurus**; **P. vulnus**.
NEMATODE, Root Knot. **Meloidogyne** sp., resistant to.
NEMATODE, Stubby Root. **Trichodorus christiei**.
NONPARASITIC. **Carapace Spot**. Abrasion of young fruits.
Dieback. Copper deficiency.
End Spot. Dessication of young fruits; overmaturity.
Little Leaf; Rosette. Zinc deficiency.
Melanorhiza. Defective drainage and aeration.
Mottle Leaf. Nutritional deficiency.
Tipburn. Sometimes salt accumulation in poorly drained soil.

Rot, Blue Mold. **Penicillium expansum**, CA.

Rot, Collar; Trunk Canker. **Phytophthora cactorum** CA; **P. parasitica** FL; Sclerotinia sclerotiorum, CA.

Rot, Fruit. **Diplodia theobromae**, FL; **Phomopsis** sp., FL, TX; **Phytophthora citrophthora**, CA; **P. citricola**, CA; **Rhizopus nigricans**; **Alternaria** sp.; **Fusarium** sp.

Rot, Postharvest Stem. **Pestalotiopsis clavispora**, Chile.

Rot, Root. **Armillaria mellea**, CA; **Clitocybe tabescens**, FL; **Phymatotrichum omnivorum**, TX; Dematophora Rot, **Rosellinia necatrix**.

Rot, Root; Decline. **Phytophthora cinnamomi**, CA; **Pythium** spp., CA.

Rot, Seed and Root. **Rhizoctonia solani**, CA.

Spot Anthracnose; Scab. **Sphaceloma perseae**, FL, TX.

Viroid. **Avocado Sunblotch**, Michoacan, Mexico.

Virus. **Avocado Sun Blotch**, CA; **Avocado 3 Alphacryptovirus**., CA, black streak.

Wilt. **Verticillium albo-atrum**, CA, FL; **Raffaelea lauricola**, FL.

Wilt, Laurel. **Raffaelea lauricola**; **R. canadensis**, CA, FL.

Wilt, Root. **Neonectria radicicola**, CA.

Avocado scab may cause heavy damage in susceptible varieties of West Indian stock. Cercospora blotch attacks both leaves and fruit. Decline, root rot, is most serious in wet soils, killing trees if they are waterlogged 6–8 days.

B

BALM (*Melissa*)
Blight, Gray Mold. **Botrytis cinerea**, AK.
Leaf Spot. **Phyllosticta decidua**, NY.

BANANA (*Musa*)
Anthracnose; Fruit Rot. **Gloeosporium musarum**, FL, TX.
Bacterial Leaf Blight. **Pseudomonas solanacearum**, FL.
Black Sigatoka. **Mycosphaerella fijiensis**, FL; Grand Bahama Island, Puerto Rico.
Fuzzy Pedicel. **Sporothrix** spp., **Fusarium pseudocircinatum**, FL.
Leaf Spot, Black. **Deightoniella**, GA.
Nematode, Burrowing. **Radopholus similis**, FL, LA.
Nematode, Lesion. **Pratylenchus musicola**, FL.
Nematode, Root Knot. **Meloidogyne** spp., FL, TX.
Rot, Root. **Clitocybe tabescens**, FL; **Ceratobasidium**, GA.
Virus. **Cucumber Mosaic**, FL.
Wilt. **Fusarium oxysporum** f. sp. **cubense**, FL.

BARBERRY (*Berberis*)
Bacterial Leaf Spot. **Pseudomonas berberidis**, general.
Blight, Gray Mold. **Botrytis cinerea**, CA, MO.
Canker; Dieback. **Botryosphaeria ribis**, FL.
Damping-Off. **Pythium debaryanum**, CA.
Leaf Spot. Gloeosporium berberidis, CT, MA, MN, OH, WI; Phyllosticta berberidis, KY.
Nematode, Root Knot. **Meloidogyne hapla**.
Nematode, Spiral. **Rotylenchus buxophilus**.
Powdery Mildew. **Erysiphe polygoni**; **Phyllactinia corylea**, MA, VT.
Rot, Heart. **Poria punctata**, MD.
Rot, Root. **Phymatotrichum omnivorum**, TX.
Rust. **Cumminsiella sanguinea** (0, I, II, III), OR; **Puccinia koeleriae** (0, I); II, III on Koehleria; **P. montanensis** (0, I); II, III on grasses.

R.K. Horst, *Field Manual of Diseases on Fruits and Vegetables*,
DOI 10.1007/978-94-007-5974-9_5, © Springer Science+Business Media Dordrecht 2013

RUST, Stripe. **Puccinia striiformis**, MN.
RUST, Wheat, **Puccinia graminis** (0, I), general; II, III on cereals and grasses.
VIRUS. **Tomato Spotted Wilt**, PA.
WILT. **Verticillium albo-atrum**, CT to VA, IL, MI.

All interstate movement of barberry is under quarantine because of the wheat rust, but resistant cultivars may be shipped under permit. Common barberry is eradicated near wheat fields; Japanese barberry is resistant to rust.

BASIL (*Ocimum*)
BACTERIAL Leaf Spot. **Pseudomonas cichorii**, LA; **P. viridiflava**, CA.
BLACK LEG. **Pleotosporium tabacinum**, IN.
DOWNY MILDEW. **Peronospora** sp., CA, FL, MA.
NEMATODE, Root Knot. **Meloidogyne** sp., FL; **M. incognita acrita**, CA.
NONPATHOGENIC. Hydroponically grown chlorosis (interveinal), CO_2 enrichment, IL.
ROT, Basal. **Rhizoctonia solani**, HI.
ROT, Root. **Phytophthora dreschleri**, NC; **Pythium irregulare**, FL.
ROT, Stem. **Sclerotinia sclerotiorum**, LA, also dieback, CA; S. minor, CA, also crown.
SPOT, Black. **Colletotrichum gloeosporioides**, FL.
VIRUS. **Bromegrass Mosaic Virus**, general, IA, SD; **Tomato Spotted Wilt**, LA, PA.
WILT. **Fusarium oxysporum** f. sp. **basilicum**, CA, FL, HI, LA, MD, MA, SC; **Sclerotinia sclerotiorum.**

BEAN, ADZUKI (*Phaseolus angularis*)
BACTERIAL Stem Rot. **Pseudomonas adzukicola**, MN.

BEAN, ASPARAGUS, YARDLONG (*Vigna sesquipedalis*)
BACTERIAL Spot. **Pseudomonas syringae**, IN, NY.
LEAF SPOT. **Mycosphaerella cruenta**, VA.
LEAF SPOT; Pod Spot. **Cladosporium vignae**, IN.
POWDERY MILDEW. **Erysiphe polygoni**, CA.
VIRUS. **Potato Virus X**, work at the University of Wisconsin on *V. unguiculata* spp. *cylindrica* "Catjang," *V. unguiculata* spp. *sesquipedalis, V. unguiculata* spp. *unguiculata*.

BEAN, FABA (*Vicia*)
VIRUS. **Bidens Mottle**, FL.

BEAN, KIDNEY, LIMA (*Phaseolus vulgaris, P. limensis*)
ANTHRACNOSE. **Colletotrichum lindemuthianum**, general in East, race 73, race 1153, ND; **C. truncatum**, PA to AL, IA, TX.
BACTERIAL Blight. **Xanthomonas phaseoli**, general in East, rare Pacific Coast.
BACTERIAL Halo Blight; Grease Spot. **Pseudomonas syringae** pv. **phaseolicola**, general in East, rare on Pacific Coast.
BACTERIAL Northern Wilt. **Clavibacter flaccumfaciens**, Northeast.
BACTERIAL Pink Seed. **Erwinia rhapontici**, Alberta, Canada.

BACTERIAL Soft Rot. **Erwinia carotovora**, cosmopolitan in market.
BACTERIAL Southern Wilt; Brown Rot. **Peudomonas solanacearum**, AL, FL, GA, OK.
BACTERIAL Spot, Leaf and Pod. **Pseudomonas syringae**, widespread.
BACTERIAL Stem Rot. **Pseudomonas adzukicola**, MN.
BACTERIAL "Stickiness". **Pseudomonas coadunata**, CA; **P. ovata**, VA.
BACTERIAL "Wildfire". **Pseudomonas syringae** pv. **tabaci**, MA, NC.
BACTERIAL Wilt. **Curtobacterium flaccumfaciens** subsp. **flaccumfaciens**, ND; Alberta, Canada, NE.
BLIGHT, Ashy Stem; Charcoal Rot. **Macrophomina phaseoli**, MD to GA, CA, CO.
BLIGHT, Gray Mold. **Botrytis cinerea**, occasional.
BLIGHT, Phytophthora. **Phytophthora capsici**. CT, IL, NY; **Choanephora cucurbitarum**, FL.
BLIGHT, Pod. **Diaporthe phaseolorum**, CT to FL, LA, OH, OK, on limabean.
BLIGHT, Southern. **Sclerotium rolfsii**, VA to FL, AR, CA, TX.
DAMPING-OFF. **Pythium ultimum**; **P. debaryanum**; **Rhizoctonia solani**.
DOWNY MILDEW. **Phytophthora phaseoli**, on lima bean, East, central states; **P. capsici**, DE, MD, NJ, VA.
LEAF AND POD SPOT. **Ascochyta boltshauseri**, NC, OR, PA; **A. phaseolorum**, WA; **Bipolaris sorokinearia**, WI.
LEAF AND STEM DISEASE. **Cercospora zebrina** (clover isolate), NC.
LEAF SPOT. **Alternaria** spp.; **Aristastoma oeconomicum**, GA, VA; **Cercospora canescens**, southeastern states to NY; **Epicoccum** sp., secondary on lima bean; **Isariopsis griseola**, Angular Leaf Spot, ME to FL, OK, TX; **I. laxa**, IN, NJ; **Mycosphaerella cruenta**, NJ to FL, AR, TX, WI; **Phyllosticta phaseolina**, NY to FL, IN, TX; **Stemphylium botryosum**; **Stagonospora phaseoli**, TN.
LEAF SPOT, White. **Pseudocercosporella albida**, MN.
LEAF SPOT, Zonate. **Cristulariella pyramidalis**, WV.
NEMATODE, Awl. **Dolichodorus heterocephalus**.
NEMATODE, Hop Cyst. **Heterodera humuli**, OR; **H. glycines**, soybean cyst, IL.
NEMATODE, Root Knot. **Meloidogyne arenaria**; **M. hapla**; **M. incognita**; **M. javanica**; Root Lesion, **Pratylenchus pratensis**.
NEMATODE, Stubby Root. **Trichodorus christiei**.
NEMATODE, Stunt. **Tylenchorhynchus claytoni**.
NONPARASITIC. **Baldhead**. Mechanical injury to seed growing point.
Blossom Drop. High temperature, low humidity.
Bronze Leaf. Excessive salt concentration, CO, MT, WY.
Chlorosis. Deficiency of: Copper, FL; Magnesium, FL, MA, SC, VA; Manganese, FL; Zinc, CA, FL. Soil alkalinity, West.
Ozone Injury. Air pollution.
Seed Pitting. Plant bug injury.
Variegation. Genetic leaf abnormality.
Wind Whip.
POD SPOT, Seed Mold. **Cladosporium herbarum**, on lima bean, CA, FL.

Pod Spot, Yeast Spot. **Nematospora phaseoli**, on lima bean; **Pullularia pullulans**, NY.

Powdery Mildew. **Erysiphe polygoni**, East, South, CA: **Microsphaera diffusa**, GA, IL, MD; **M. euphorbiae**, IL.

Rot, Black Root. **Thielaviopsis basicola**, AL, CA, NH, NJ, NY.

Rot, Brown Stem. **Cephalosporium gregatum**, IL.

Rot, Root. **Fusarium solani** f. sp. **phaseoli**, general; **F. acuminatum, F. reddens**, MN, ND; **Phoma terrestris**, secondary, CA; **Phymatotrichum omnivorum**, AZ, OK; **Pythium aristorum**, WI; **P. catenulatum**, WI; **P. dissotocum**, WI; **Pythium** spp., widespread; **Aphanomyces euticheses**, NY.

Rot, Root and Stem; Web Blight. **Pellicularia filamentosa**, general.

Rot, Soft. **Rhizopus stolonifer** and **Pythium aphanidermatum**, leak, of market beans.

Rot, Stem; Wilt. **Sclerotinia sclerotiorum**, general, especially in South and West; **S. minor**; **S. ricini**; **S. intermedia**, in market beans. **Pythium myriotylum**, FL, GA, SC; **Marasmius** spp., NE.

Russet. **Plectosporium tabacinum**, MD, NY.

Rust. **Uromyces phaseoli** var. **typica**, II, III, general; 0, I, rare; **U. appendiculatus** var. **appendiculatus** (= **U. phaseoli**), CO, NE; **Phakopsora pachyrhizi**, FL.

Stem Necrosis, Decline. **Phytophthora capsici**, MI.

Spot Anthracnose; Lima Bean Scab. **Sphaceloma phaseoli**, NC.

Virus. **Alfalfa Mosaic**, WA; **Bean Common Mosaic**, ID, NY, WA; **Bean Common Mosaic Necrosis**, CA; **Bean Mosaic**, general; **Southern Bean Mosaic**, CA, CO, LA, MD, MI; **Bean Pod Mottle**; **Bean Phyllody**; **Bean Red Node**; CO, ID, MS, WY; **Bean Yellow Mosaic**, ND; **Bean Yellow Stipple**, IL; **Bean Yellow Dot = Alfalfa Mosaic**, WA; **Beet Curly Top**, West; **Cucumber Mosaic**; **Tobacco Necrosis**; **Tobacco Ring Spot**; **Bean Golden Mosaic**, FL; **Cowpea Chlorotic Mottle**, IL; **Clover Proliferation Subgroup A Phytoplasma**, WA; **Bromegrass Mosaic**, general, IA, SD, on Broad Bean (*Vicia fava* var. *minor*); **Clover Yellow Vein**, CA; **Cucurbit Yellow Stunting Disorder**, CA; **Lettuce Mosaic**, NY; **Lima Bean Mild Mottle**, GA; **Potato Virus X**, work at the University of Wisconsin on *P. lathyroides, P. lunatus, P. aboriginens, P. vulgaris* "Pinto"; **Peanut Stunt**, AR; **Peanut Mottle**, WA; **Cucurbit Leaf Crumple**, FL; **Sida Golden Mosaic**, FL; **Sida Micrantha Mosaic**, Brazil.

Wilt; Yellows. **Fusarium oxysporum** f. sp. **phaseoli**, CA, CO, ID, MT; **F. oxysporum** f. sp. **vasinfectum**, AL, FL.

Anthracnose and bacterial blight, often erroneously called "rust," are common and destructive bean diseases best avoided by purchasing healthy seed grown in disease-free arid sections of California and the Northwest. True rust is prevalent in the Southwest and sometimes in the East on susceptible Kentucky Wonder pole beans. Use resistant varieties; clean and disinfest poles each season.

Downy mildew is common on lima beans in moist summers. Resistant varieties are the best solution to virus problems. Avoid picking or cultivating beans when foliage is wet.

BEAN, MUNG (*Vigna radiata*)
DAMPING-OFF. **Rhizoctonia solani**, CA.
ROT, Sprout. **Cylindrocephalum** sp., WA.

BEAN, SCARLET RUNNER (*Phaseolus coccineus*)
ANTHRACNOSE. **Colletotrichum lindemuthianum**, NY.
BACTERIAL Blight. **Xanthomonas phaseoli**, IN, NJ, TX.
BACTERIAL Halo Spot. **Pseudomonas phaseolicola**, NY.
LEAF SPOT. **Ascochyta boltshauseri**, OR; **Cercospora cruenta**, AL.
POWDERY MILDEW. **Erysiphe polygoni**, CA.
ROT, Root. **Fusarium solani** f. sp. **phaseoli**, NY.
RUST. **Uromyces phaseoli** (II, III), AL, MA, NH; **Phakopsora pachyrhizi**, FL.

BEAN, TEPARY (*Phaseolas acutifolius*)
BLIGHT, Southern. **Sclerotium rolfsii**, AL.
POWDERY MILDEW. **Erysiphe polygoni**, CA.
ROT, Root. **Fusarium solani**, CA; **Phymatotrichum omnivorum**, TX.
RUST. **Uromyces phaseoli** (II, III), CA, TX.
VIRUS. **Beet Curly Top**, CA.

BEAN, URD (*Vigna*)
BACTERIAL Leaf Spot; Stem Collapse. **Pseudomonas syringae** pv. **syringae**, ID.
VIRUS. **Cowpea Mosaic**, NY.

BEAN, WINGED (*Psophocarpus tetragonolobus*)
VIRUS. **Cucumber Mosaic**, FL; **Clover Yellow Vein**, MD.

BEET (*Beta vulgaris*)
BACTERIAL, symtomless. **Clavibacter sepedonicum**, ND.
BACTERIAL Black Streak. **Pseudomonas syringae** pv. **aptata**, CA, OR, UT, WA.
BACTERIAL Crown Gall. **Agrobacterium tumefaciens**, occasional.
BLIGHT, Foliage. **Thanatephorus cucumeris**, MN.
BLIGHT, Phytophthora. **Phytophthora**, IL.
BLIGHT, Southern. **Sclerotium rolfsii**, NC to FL, AZ, CA, TX.
DAMPING-OFF. **Pythium** spp.; **Rhizoctonia solani**, general.
DODDER. **Cuscuta** spp., occasional when beets follow legumes.
DOWNY MILDEW. **Peronospora schachtii**, CA, MN, NJ, NY, OR, WA.
LEAF SPOT. **Alternaria** sp., general, secondary; **Gloeosporium betae**, MS, MT; **Ramularia beticola**, CA, OR, WA; **Septoria betae**, DE, IN, MA, OH.
LEAF SPOT; Blight. **Cercospora beticola**, general.
NEMATODE, Leaf and Stem. **Ditylenchus dipsaci**, KS.
NEMATODE, Root Knot. **Meloidogyne arenaria**; **M. javanica**; **M. chitwoodi**, Pacific NW; **M. hapla**, Pacific NW; Sugar-beet, **Heterodera schachtii**; Clover Cyst, **H. trifolii**; Gall, **Nacobbis Batatiformis**.
NEMATODE, Sheath. **Hemicycliophora obtusa**. Sting, **Belonolaimus gracilis**; Stubby Root, **Trichodorus christiei**; Lesion, **Pratylenchus penetrans**.
NONPARASITIC. **Black Heart**. Boron deficiency, general; phosphorus deficiency, occasional.

Bronzing. Potassium deficiency occasional.
Chlorosis. Alkalinity, iron, manganese deficiency.
Girdle. Strangling constriction of tap root.
Tipburn. Black tips, from high nitrogen with low light intensity.

POWDERY MILDEW. **Erysiphe polygoni**, CA, MI.
ROT, Black Heart. **Phoma betae**, general, occasional as leaf spot.
ROT, Black Root. **Aphanomyces cochlioides**, general.
ROT, Charcoal. **Macrophomina phaseoli**, CA.
ROT, Crown. **Sclerotinia sclerotiorum**, CT, IL.
ROT, Root. **Phytophthora drechsleri**, CA, CO, ID; **P. cryptogea**, WY; **Physalospora rhodina**, AL; **Phymatotrichum omnivorum**, TX; **Pythium deliense**, AZ, TX; **Rhizopus arrhizus**, WY; **Fusarium oxysporum** f. sp. **betae**, CO, MT.
ROT, Violet Root. **Helicobasidium purpureum**, occasional.
ROT, Wound, Storage. **Aspergillus fumigatus**, MI; **Rhizopus** spp., cosmopolitan; **Cylindrocarpon radicicola**, NY; **Penicillium** spp.; **Fusarium** spp.
RUST. **Puccinia aristidae** (0, I), CO, KS, NM, UT; II, III on grasses; **P. subnitens**, NE; **Uromyces betae** (II, III), AZ, CA, NM, OR.
SCAB. **Streptomyces scabies**, widespread.
SCAB, Acid. **Streptomyces acidiscabies**, ME.
VIRUS. **Beet Black Scorch**, CO; **Beet Curly Top**; **Beet Latent**; **Beet Mosaic**; **Beet Pseudo- Yellows**; **Beet Ring Mottle**; **Beet Savoy**; **Beet Yellow Net**; **Beet Yellows**; **Beet Yellow Vein**; **Cucumber Mosaic**; **Beet Necrotic Yellow Vein**, CA, NE, TX, WA; **Beet Necrotic Yellow Vein** (Vector, *Polymyxa beta*, **Rhizomania** disease), CA, MN, TX, WA; **Beet Western Yellows**, CA; **Lettuce chlorosis**, CA; **Lettuce Infectious Yellows**, AZ, CA; **Beet Distortion Mosaic**, CA; **Beet 2 Alphacryptovirus**; **Lettuce Speckles Mottle**, CA; **Beet Soilborne**; **Beet Soilborne Mosaic**, MN; **Curtoviruses**, AZ.
WHITE RUST. **Albugo bliti**, IA, OH.
WILT. **Verticillium albo-atrum**, CO.
WILT; Yellows. **Fusarium** sp., MN; **Fusarium conglutinans** (**F. orthoceras**) var. **betae**; **F. graminearum** (sugar beets) MN, WY.

Most of the troubles are more important for commercially grown sugar beets. Cercospora leaf spot or blight is common on garden beets although spraying is not often practical. Boron deficiency can be prevented by treating soil with borax.

BLACKBERRY (*Rubus*)

ANTHRACNOSE. **Glomerella cingulata**, Mexico.
BACTERIAL Crown Gall. **Agrobacterium tumefaciens**, general; Cane Gall, **A. rubi**, NY, OR, PA, WA, WI; Hairy Root, **A. rhizogenes**, OR.
BLIGHT, Cane. **Gnomonia rubi**, ME, MD, NY, PA, VT; **Leptosphaeria coniothyrium**, NY to NC, TX, WI, Pacific Northwest.
BLIGHT, Leaf. **Colletotrichum acutatum**, Brazil.
BLIGHT, Spur. **Didymella applanata**, VA, WI.
BLIGHT, Stamen; Dry Berry. **Hapalosphaeria deformans**, OR, WA.
BLIGHT, Thread. **Pellicularia koleroga**, LA.

BLOTCH, Sooty. **Gloeodes pomigena**, MD to NC, IN, TX; **Geastrumia polystigmatis**, NC.
CANKER. **Glomerella cingulata**, MD, VA; **Phomopsis** sp., WA; **Botryosphaeria obtusa**, OH; **B. dothidea**, MD; **Leptosphaeria coniothyrium**, OH; **Gnomonia rubi**, OH.
CANKER, Cane Spot. **Ascospora ruborum**, WA.
DOWNY MILDEW. **Peronospora rubi**, MD, WI; **P. sparsa**, Mexico.
FRUIT SPOT; Flyspeck. **Leptothyrium pomi**, PA to NC, IL.
LEAF SPOT. **Mycosphaerella confusa**, VA to FL, IL, IN, TX; **M. rubi**, NC, WI; **Pezizella oenotherae**, also fruit rot, MD, OH, VA; **Cylindrosporium rubi**, NC, TX; **Phyllosticta** spp., FL, IL, NH; **Septoria darrowi**; **S. rubi**, general.
LEAF SPOT. Algal. **Cephaleuros virescens**, AR, FL, LA.
POWDERY MILDEW. **Sphaerotheca macularis**, CT to MD, IL, MN, Pacific Northwest; **Phyllactinia corylea**, MI.
ROSETTE, Double Blossom. **Cercosporella rubi**, NY to FL, CA, TX.
ROT. **Botrytis** spp., SC.
ROT, Gray Mold. **Botrytis cinerea**, general on fruit, bud, shoot.
ROT, Root. **Armillaria mellea**, TX, WA; **Phymatotrichum omnivorum**, TX; **Phytophthora dreschleri**, NC; **Rhizoctonia solani**.
ROT, White Root. **Corticium galactinum**, near apple trees, AR, MD, VA.
RUST. **Phragmidium violaceum**, CA.
RUST, Orange. **Gymnoconia peckiana** (0, I, III), ME to GA and west to Pacific; **Kunkelia nitens** (I), general, more common in South; **Phragmidium violaceum**.
RUST, Yellow Cane. **Kuehneola uredinis** (0, I, II, III), ME to FL, LA, WI.
SPOT ANTHRACNOSE. **Elsinoë veneta**, general.
VIRUS. **Blackberry Dwarfing**; **Blackberry Mosaic**; **Blackberry Variegation**; **Loganberry Dwarf**, OR; **Raspberry Beta-Leaf Curl**; **Red Raspberry Mosaic**, MA to VA, IA, WI, Pacific Northwest; **Raspberry Ringspot**; **Raspberry Streak**, OH, PA, WA; **Tobacco Streak**, OR, WA; **Wineberry Latent**; **Beet Pseudo-Yellows**, AR, NC, SC; **Marafivirus**, MS; **Impatiens Necrotic Spot**, Southern U.S, AL; **Tobacco Ringspot**, **Tomato Ringspot**, **Raspberry Bushy Dwarf**, AL; **Black Raspberry Necrosis**, TN; **Blackberry Chlorotic Ringspot**, OR, WA.
WILT. **Verticillium albo-atrum**, CA, MN, NY, WA.

Sanitation is the best approach to home garden blackberry diseases. Plants with crown gall, orange rust, or virus diseases should be removed and burned, replanting with clean stock.

BLUEBERRY (*Vaccinium*)
ANTHRACNOSE. **Gloeosporium** sp., FL, NJ.
BACTERIAL Canker. **Pseudomonas** sp., OR; **Xanthomonas campestris** pv. **manihotis** (also leaf spot), FL; **X. campestris** pv. **fici** (also leaf spot), FL.
BACTERIAL Crown Gall. **Agrobacterium tumefaciens**, MA, MI, WA.
BACTERIAL Leaf Scorch. **Xylella fastidiosa**, FL.

Black Mildew. **Meliola nidulans**, AL, GA.

Blight, Blossom, Fruit, Twig. **Botrytis cinerea**, CA, ME, MA, NJ, OR, WA; **Diaporthe vaccinii**, ME, MA, NJ.

Blight, Stem. **Botryosphaeria dothidea** (*B. ribis*), NC; **Phomopsis vaccinii**, NC; **Lasiodiplodia theobromae,** FL; **Fusarium** sp., Argentina; **Calonectria colhounii**, MI.

Blight, Twig. **Nigrospora sphaerica**, Argentina.

Canker, Blight of Stem, MI. **Glomerella cingulata**, **Alternaria** sp., **Sordaria** sp., **Epicoccum** sp., **Tympanis** sp., **Papulospora** sp., **Nectria cinnabarina**, **Coniothyrium** sp., **Pestalotia** sp., **Verticillium** sp., **Godronia cassandrae** f. sp. **vaccinii**, **Phoma** sp., **Dendrophoma** sp., **Diaporthe** (**Phomopsis**) **vaccinii**, **Melanospora** sp., **Pullularia** spp., **Coryneum microstictum**, **Sphaeronema** sp., **Fusarium** spp., **Bispora** sp., **Botrytis cinerea**, **Cephalosporium** sp., **Cylindrocarpon** sp., **Pyrenochaeta** sp.

Canker, Cane. **Physalospora corticis**, AL, FL, MS, NC; **Godronia cassandrae**, ME, MI; **Botryosphaeria corticis**, AL, FL, GA, MS, NC; **Coryneum microstictum**, MA, MI.

Canker, Twig. **Dothidella vacciniicola**, NC; **Fusicoccum putrefaciens**, ME, MA, MI, **Phomopsis vaccinii**, MI; **Sporocadus lichenicola**, OR.

Dodder. **Cuscuta** sp., FL.

Gall, Crown and Root. **Agrobacterium rubi** and **A. rhizogenes**, Argentina.

Gall, Red Leaf. **Exobasidium vaccinii**, also fruit green spot, general; **Synchytrium vaccinii**, ME, MS.

Leaf Rust. **Thekopsora minima**, MI, Mexico; **Pucciniastrum corymbosum**, HI.

Leaf Spot. **Gloeocereocospora inconspicua**, GA, MD, NC; **Phyllosticta** sp., AL, GA; **Phyllostictina vaccinii**, GA, MD, MI, NC; **Piggotia vaccinii**, WI; **Discohainesia oenotherae**, NJ, NC; **Ramularia vaccinii**, FL, MD, MI, NJ, NY; **Septoria albopunctata**; **Alternaria tenuissima**, NC; **Cristulariella pyramudates**, MD; **Nigrospora sphaerica**, Argentina.

Leaf Spot, Double Spot. **Dothichiza caroliniana**, GA, NC.

Leaf Spot, Tar. **Thytisma vaccinii**, widespread.

Nematode. **Helicotylenchus** sp.; **Hemicycliophora similis**; **Merlinius joctus**; **Xiphinema Americanum**; **Paratrichodorus renifer**, WA, British Columbia, Canada.

Powdery Mildew. **Microsphaera alni** var. **vaccinii**, widespread.

Rot, Berry; Dieback. **Alternaria** sp., MA, NJ, NC; **Glomerella cingulata**, NJ, NY; **Phomopsis vaccinii**, NC.

Rot, Brown; Mummy Berry. **Monilinia vaccinii-corymbosi**, IN, ME, MA, MI, NJ, NY, NC, PA; **Sclerotinia sclerotiorum**, Argentina.

Rot, Root. **Phytophthora cinnamomi**, AR, NJ, VA; **P. nicotainae**, VA; **P. palmivora**, VA; **Fusarium proliferatum**, Argentinia.

Rust, Witches' Broom. **Pucciniastrum goeppertianum** (III) ME to MN; 0, I on fir; Leaf, **P. myrtilli** (II, III) ME to PA, WI; 0, I, spruce.

Virus. **Blueberry Necrotic Ring Spot**; **Blueberry Red Ring Spot**; **Blueberry Stunt**, MA, MI, NJ, NY, NC; **Blueberry Shoestring**, MI, NJ, WA; **Blueberry**

Leaf Mottle, MI; **Blueberry Necrotic Shock**, OR, WA; **Blueberry Scorch**, CT, MA, OR, WA, British Columbia, Canada; **Tobacco Ring Spot**, IL; **Peach Rosette Mosaic**, MI.

BOYSENBERRY

Subject to most blackberry diseases: anthracnose, crown gall, cane gall, cane canker, dieback, leaf spots, mosaic.

BROCCOLI
See Cabbage.

BRUSSEL SPROUTS
See Cabbage.

CABBAGE (*Brassica oleracea*) (Including Broccoli, Brussel Sprouts, Cauliflower, Kale, Kohl-rabi)
BACTERIAL Black Rot. **Xanthomonas campestris**, general.
BACTERIAL Blight. **Pseudomonas syringae**, CA; **P. syringae** pv. **alisalensis**, CA, SC; **P. cannabina** pv. **alisalensis**, CA.
BACTERIAL Leaf Spot. **Pseudomonas maculicola**, widespread; **P. cichorii**, FL and in market; **Xanthomonas campestris** pv. **armoraciae**, FL, OH; **X. campestris** pv. **campestris,** SC; **Pseudomonas syringae** pv. **maculicola**, CA, OH, SC.
BACTERIAL, MLO. Western **Aster Yellows Phytoplasma**, CA, TX.
BACTERIAL Soft Rot; Stump Rot; **Erwinia carotovora**, cosmopolitan.
BLACKLEG. **Leptosphaeria biglobosa**, Central States, Mexico.
BLACKLEG; Leaf Spot. **Phoma lingam**, general east of Rocky Mts., also OR, WA.
BLIGHT, Gray Mold. **Botrytis cinerea**, FL; Pacific states, frequent.
BLIGHT, Southern. **Sclerotium rolfsii**, NC to FL, TX, occasional.
CANKER, Stem. **Botrytis cinerea**, WA.
CLUB ROOT. **Plasmodiophora brassicae**, general.
DAMPING-OFF; Bottom Rot. **Pythium** spp.; **P. polymastum**, CA; **Rhizoctonia solani**, general.
DOWNY MILDEW. **Peronospora parasitica**, general.
LEAF SPOT, Gray. **Alternaria brassicae**, general; Black Spot, Head Browning, **A. brassicicola**, general; **A. oleraceae**, occasional; **Cercospora brassicola**, CA, DE, FL, IL, MS, NC, OK; **Phyllosticta brassicicola**, CA; **Leptosphaeria maculans**, WI.
LEAF SPOT; White Spot. **Cercosporella brassicae**, IN, OR, PA; Ring Spot, **Mycosphaerella brassicicola**, AL, CA, IL, NY, OR, TX, WA; **Pseudocercosporella capsellae**, CA.
MOLD, Leaf. **Heterosporium variabile**, MT, NY; Seed, **Curvularia geniculata**; **Stemphylium botryosum**.
NEMATODE. **Nacobbis batatiformis**.
NEMATODE, Cabbage Cyst. **Heterodera cruciferae**. Sugar-Beet, **H. schachtii**.
NEMATODE, Lesion. **Pratylenchus pratensis**. Sting, **Belonolaimus gracilis**; **B. longicaudatus**; Stubby Root, **Trichodorus christiei**.

NEMATODE, Root Knot. **Meloidogyne arenaria**; **M. javanica**.
NONPARASITIC. **Brown Heart**. Probably boron deficiency.
  Chlorosis. Magnesium or manganese deficiency.
  Oedema. Excessive water tension, or copper sprays.
  Pink Head. Probably genetic.
  Tipburn. Potassium deficiency.
  Whip-tail. Spindly growth, failure to head, due to acid soil, mineral deficiencies.
POWDERY Mildew. **Erysiphe polygoni**, AZ, CA, CT, FL, MD, MA, OR; **E. cruciferarum**, CA.
ROT, Cottony; Drop. **Sclerotinia sclerotiorum**, AZ, MA, NY, TX.
ROT, Firm Head. **Sclerotinia sclerotiorum**, CA, NY; **Phytophthora porri**, WI.
ROT, Head. **Fusarium avenaceum**, NY.
ROT, Root. **Macrophomina phaseoli**, CA. on *Brassica campestris*; **Phymatotrichum**, TX; **Phytophthora megasperma**, CA, OR; **P. drechsleri**, KY.
ROT, Root and Crown. **Sclerotinia minor**, CA.
ROT, Root; Damping-off. **Aphanomyces raphani**, WI. Also: **Brassica alboglabra**–Chinese Kale; **B. carinata**–Abyssinian Mustard; **B. napus**–; Rape; **B. perviridis**–Spanish Mustard; **B. robertiana**; **Eruca sativa**–Racuet Salad; **Raphanus raphanistruno**–Wild Radish.
ROT, Soft; Black Mold. **Rhizopus stolonifer**, occasional.
ROT, Sprout. **Alternaria tenuis**, **Rhizoctonia solani**, and **Fusarium** sp. on Brussels sprouts.
VIRUS. **Cabbage Ring Necrosis**; **Cauliflower Mosaic**, GA; **TurnipMosaic**, GA; **Beet Curly Top**; **Tomato Spotted Wilt**; **Cabbage Leaf Curl**, GA.
WHITE RUST; Blister. **Albugo candida**, CA, ID, KY, NE, NC, OH, TX.
WILT, Vascular. **Verticillium dahliae**, CA.
WILT, Yellows. **Fusarium oxysporum** f. sp. **conglutinans**, general.

A general control program starts with choosing varieties resistant to Fusarium yellows, purchasing disease-free seed or having it hot-water treated against blackleg and black rot, growing in soil free from club root and careful cleaning up of all vegetable refuse at the end of the season.

C

CANTALOUPE
See Melon.

CARAWAY (*Carum*)
Bacterial, MLO. **California Aster Yellows**, WA.
Dodder. **Cuscuta** sp., WA.
Nematode, Root Knot. **Meloidogyne** sp.
Rot, Stem. **Sclerotinia sclerotiorum**, WA.

CARDOON (*Cynara cardunculus*)
Leaf Spot. **Cercospora obscura**, CA, TX; **Ramularia cynarae**, CA.
Powdery Mildew. **Golovinomyces cichoracearum**, CA.

CARROT (*Daucus carota* var. *sativa*)
Bacterial Blight. **Xanthomonas carotae**, CA, ID, IA, WI.
Bacterial Core-Rot. **Erwinia chrysanthemi** and/or **E. carotovora** var. **carotovora**, IL.
Bacterial, MLO. **Aster Yellows**, widespread and California strain, CA, ID, OR, WA.
Bacterial Soft Rot. **Erwinia carotovora**, general.
Blight, Early; Leaf Spot. **Cercospora carotae**, general.
Blight, Late. **Alternaria dauci**, general.
Blight, Southern. **Sclerotium rolfsii**, GA to FL, CA, TX.
Canker, Root; Storage Rot. **Rhizoctonia** spp., MI, NY, OR, WA.
Damping-Off. **Pythium** sp., ID; **Rhizoctonia solani**, general.
Dodder. **Cuscuta** sp., ID, MS, NM, NY, TX, WV.
Leaf Spot. **Ramularia** sp., KS; **Alternaria tenuis**, secondary, also seed mold; **Cercospora carotae**, WA.
Nematode. **Nacobbis batatiformis**.
Nematode, Cyst. **Heterodera carotae**.
Nematode, Root Knot. **Meloidogyne arenaria**; **M. incognita**; **M. javanica**.
Nonparasitic. **Black Heart**. Cause unknown, WI.

R.K. Horst, *Field Manual of Diseases on Fruits and Vegetables*,
DOI 10.1007/978-94-007-5974-9_6,

Chlorosis. Magnesium deficiency in acid soil.
Root Girdle. Cause unknown.

POWDERY MILDEW. **Erysiphe polygoni**, CA; **E. heraclei**, NY, NC, WA, Michoacan, Mexico.

ROT, Black. **Alternaria radicina**, ID, MA, MI, NY, PA, WA.

ROT, Black Mold. **Rhizopus** spp.; Blue Mold, **Penicillium** sp.; Gray Mold, **Botrytis cinerea**; Pink Mold, **Trichothecium roseum**, IN.

ROT, Crown. **Rhizoctonia solani**, GA.

ROT, Dry. **Fusarium** spp., ID, NY, associated with scab; **Mycocentrospora acerina**, WA.

ROT, Root. **Helicobasidium purpureum**, OR, WA; **Phymatotrichum omnivorum**, AZ, LA, TX; **Phytophthora megasperma**, CA; **P. cactorum**, MI; **Sclerotinia minor**, **S. sclerotiorum**, Ontario, Canada.

ROT, Storage. **Centrospora acerina**, NY; **Aspergillus niger**; **Pellicularia filamentosa**; **Typhula** sp.

ROT, Watery Soft. **Sclerotinia sclerotiorum**, general; **S. intermedia**, **S. minor**.

RUST. **Uromyces scirpi** (0, I), OR; II, III on *Scirpus*.

SCAB. **Streptomyces scabies**, CA, MI, PA, WA.

SCAB, Acid. **Streptomyces acidiscabies**, ME.

SPIROPLASMA. **Carrot Purple Leaf**, WA.

VIRUS. **Alfalfa Mosaic**, CA, WA; **Clover Yellow Vein**, WA; **Beet Curly Top**, OR, UT; **Carrot Motley Dwarf**; **Celery Mosaic**; **Carrot Red Leaf**; **Carrot Thin Leaf**, CA; **Lettuce Infectious Yellows**, AZ, CA.

Carrots in home gardens require a deeply dug friable soil even more than treatment for diseases. Root-knot nematodes are rather common.

CASHEW (*Anacardium*)

BLIGHT, Seedling. **Sclerotium rolfsii**, FL.

CELERY, CELERIAC (*Apium graveolens*)

BACTERIAL Leaf Spot. **Pseudomonas jaggeri** pv. **apii**, CT, DE, FL, IN, MI, MN, ND, NY, OH; Leaf Blight, **P. cichorii**, FL; **Pseudomonas syringae** pv. **apii**, CA.

BACTERIAL, MLO. **California Aster Yellows**, CA, ID, UT, WA.

BACTERIAL Soft Rot. **Erwinia carotovora**, cosmopolitan.

BLIGHT, Early. **Cercospora apii**, general.

BLIGHT, Late. **Septoria apiicola** (includes *S. apii* large leaf spot, and *S. apii-graveolentis*), general.

DAMPING-OFF. **Pythium mastophorum**, CA; **Pythium** spp.; **Rhizoctonia solani**; **Aphanomyces euteiches**, cosmopolitan.

LEAF SPOT. **Phyllosticta apii**, DE, NJ; **Stemphylium** sp. and **Alternaria** sp., secondary.

NEMATODE, Awl. **Dolichodorus heterocephalus**, red root, FL.

NEMATODE, Pin. **Paratylenchus hamatus**, New England, CA; **P. projectus**, NJ.

NEMATODE, Root Knot. **Meloidogyne incognita**.

NEMATODE, Sheath. **Hemicycliophora parvana**, FL.
NEMATODE, Stem. **Ditylenchus dipsaci**, CA.
NEMATODE, Sting. **Belonolaimus gracilis**, FL; **B. longicaudatus**.
NEMATODE, Stubby Root. **Paratrichodorus christiei**.
NONPARASITIC. **Black Heart**. Wide fluctuations of soil moisture, general.
Brown Stem. Over-age plants.
Cracked Stem. Boron deficiency.
Hollow Stem. Pithiness. Sometimes chilling; sometimes genetic.
POWDERY MILDEW. **Erysiphe heraclei**, CA.
ROT, Gray Mold. **Botrytis cinerea**, cosmopolitan.
ROT, Petiole and Crown. **Sclerotinia minor**, CA.
ROT, Pink; Watery Soft. Sclerotinia sclerotiorum, general; S. intermedia; S. minor.
ROT, Root; Scab on celeriac. **Phoma apiicola**, CA, MI, NY, OH, WI.
ROT, Stem; Brown Spot. **Cephalosporium apii**, CA.
ROT, Storage. **Centrospora acerina**, NY; **Typhula variabilis**, NY; **Tricothecium roseum**, pink mold, NY.
SLIME MOLD. **Physarum** spp., CA.
VIRUS. **Celery Calico**, CA, FL, NY, OH; **Celery Mosaic**, AL, CA, FL, WA; **Cucumber Mosaic**; **Beet Curly Top**, OR; **Celery Yellow Spot,** CA; **Tomato Spotted Wilt**; **Lettuce Mosaic**, NY; **Broad Bean Wilt**, NY; **Apium Virus Y,** CA, WA; **Potyvirus**, CA.
WILT, Yellows. **Fusarium oxysporum** f. sp. **apii**, CA, MI, NY, TX, general.
WILT. **Verticillium albo-atrum**, CA.

In many areas soil treatment for nematodes increases yield. Yellows-resistant varieties are on the market. Seed should be treated for leaf blights unless more than 2 years old. In Florida, development of apothecia of *Sclerotinia* is inhibited by flooding fields. Western growers fight mosaic by celery-free periods and by controlling insect vectors.

CHAYOTE (*Sechium*)
ANTHRACNOSE. **Colletotrichum lagenarium**, FL, TX.
BLIGHT, Southern. **Sclerotium rolfsii**, TX.
LEAF SPOT. **Cercospora sechii**, FL, TX.
NEMATODE, Root Knot. **Meloidogyne** sp.; **M. incognita**.
NEMATODE, Stubby Root. **Trichodorus** sp.
POWDERY MILDEW. **Podosphaera xanthii**, LA.
ROT, Fruit. **Glomerella cingulata**, LA.

CHERRY (*Prunus* spp.)
BACTERIAL Black Spot; Canker; Gummosis. **Xanthomonas pruni**, NY to MI, GA and TX.
BACTERIAL Canker. **Pseudomonas syringae**, MA to MI; Pacific states and **P. morsprunorum**.
BACTERIAL Crown Gall. **Agrobacterium tumefaciens**, CA, TX, WA.
BACTERIAL Fire Blight. **Erwinia amylovora**, OR, WA.

Bacterial, MLO. **Peach X-Disease**; **Peach Yellow Leaf Roll** (**Peach Western X-Disease**), Northwest.

Black Knot. **Dibotryon morbosum**, eastern states.

Blight, Blossom, Brown Rot. **Monilinia laxa**, general, Pacific states, MI, NY, WI; **Sclerotinia sclerotiorum,** OR.

Blight, Seedling, Twig. **Monilinia rhododendri**, VT to GA, AR, IA; **M. fructicola**, shoot and leaf, GA.

Blight, Shoot; Shot Hole. **Coryneum carpophilum** (*C. beijerinckii*), CA,ID, OR, WA; **Monilinia laxa**; MA, NY, RI.

Canker, Felt Fungus. **Septobasidium retiforme**, GA.

Canker, Trunk and Collar. **Phytophthora cactorum**, **P. citrophthora**, CA.

Canker, Twig. **Phomopsis padina**; **Cytospora leucostoma**, NY, PA; **Valsa leucostoma**, widespread; **Nectria** sp., NY; **Eutypa lata** (syn. **E. armeniacae**), CA.

Dieback, Branch. **Calosphaeria pulchella**, CA.

Leaf Blister. **Taphrina farlowii**, VT to FL, TX; **Taphrina cerasi**, ME to NJ, MN.

Leaf Spot. **Cercospora circumscissa** (*Mycosphaerella cerasella*), NJ, PA, VA to FL, TX; **Alternaria citri** var. **cerasi**, CA; **Phyllosticta pruni-avium**, OR; **Cercospora graphioides** and **Phyllosticta serotina** on wild black cherry only; **Coccomyces lutescens**, VA.

Leaf Spot; Blight; Shot hole. **Coccomyces hiemalis**, general.

Mistletoe. **Phoradendron serotinum** (**flavescens**).

Nematode, Lesion. **Pratylenchus** sp.; **P. vulnus**, CA.

Nematode, Root Knot. **Meloidogyne** sp; **M. incognita**; **M. javanica**.

Nonparasitic. **Brown Bark Spot**. Arsenical poisoning of soil, ID, MT, WA. **Chlorosis**. Alkali injury, CA, TX. **Little Leaf**. Zinc deficiency, CA, OR, WA.

Powdery Mildew. **Podosphaera oxyacanthae**, general.

Rot, Brown. **Monilinia fructicola**, general; **M. laxa**, MA, NY, RI, Argentina.

Rot, Fruit. **Alternaria** sp.; **Botrytis cinerea; Cladosporium herbarum**, CA to WA and ID; **Lambertella** sp., OR; **Microstroma tonellianum**, MA; **Penicillium expansum; Rhizopus stolonifer**, occasional in market; **Pullularia** sp., Northwest; **Mucor piriformis.**

Rot, Green Fruit. **Sclerotinia sclerotiorum**, OR.

Rot, Heart; Wood. **Fomes fomentarius**, Northeast; **Polyporus** spp.; **Poria ambigua**, CA.

Rot, Root. **Armillaria mellea**, NM, OK, OR; **Phymatotrichum omnivorum**, TX; **Xylaria** spp., VA; **Poria ambigua**, CA.

Rot, Silver Leaf. **Stereum purpureum**, NY, MT.

Rust. **Tranzschelia discolor** (II, III), GA, MA, NE, NY, NC, OK, TX; 0, I on anemone.

Scab. **Fusicladium cerasi**, NY to IA and WI.

Virus. **Cherry Rough Fruit**, UT on sweet cherry (*Prunus avium*); **Cherry Albino**; **Cherry Bark-Splitting**; **Cherry Black Canker; Cherry Buckskin**; **Cherry Bud Abortion**; **Cherry Chlorosis**; **Cherry Freckle Fruit Disease**; **Cherry Green**

Ring Mottle; **Cherry Gummosis**; **Cherry Little Cherry**; **Cherry Midleaf Necrosis**; **Cherry Mora**; **Cherry Mottle Leaf**; **Cherry Necrotic Rusty Mottle**; **Cherry Pink Fruit**; **Cherry Pinto Leaf**; **Cherry Rasp Leaf**; **Cherry Ring Spot**; **Cherry Rugose Mosaic**; **Cherry Rusty Mottle**; **Cherry Twisted Leaf**; **Cherry Vein Clearing**; **Cherry Yellows**; **Prune Dwarf**; **Tomato Ring Spot**; **Stem Pitting**; **Prunus Necrotic Ring Spot,** CA, WA; **Plum Pox**, WA.
VIRUS, Leafroll and Yellow Leaves. **Nepovirus**, WA.

CHERRY-LAUREL (*Prunus laurocerasus*)
BACTERIAL Spot. **Xanthomonas pruni**, GA, MS, NJ, SC.
BLIGHT, Blossom; Brown Rot. **Monilinia fructicola**, CA; **M. laxa**, CA; **Alternaria** sp., TX.
LEAF SPOT. **Cercospora circumscissa**, CA; **C. cladosporioides**, LA, TX; **Coccomyces lutescens**, MS; **Phyllachora beaumontii**, AL; **Phyllosticta laurocerasi**, CA, FL, NJ; **Septoria ravenelii**, SC.
MISTLETOE. **Phoradendron serotinum** (**flavescens**), FL.
ROT, Root. **Armillaria mellea**, CA; **Clitocybe tabescens**, FL; **Phymatotrichum omnivorum**, TX.
WILT. **Verticillium albo-atrum**, CA.

CHICKORY
See Endive.

CHICK-PEA, GARBANZO (*Cicer*)
BACTERIAL, Blight. **Pseudomonas andropogonis**, ME.
BACTERIAL, Pink Seed. **Erwinia rhapontici**, Saskatchewan, Canada.
BLIGHT. **Ascochyta radiei**, ID, NE, WA.
NEMATODE, Root Knot. **Meloidogyne** spp.
POWDERY MILDEW. **Leveilula taurica**, WA.
ROT, Root. **Fusarium solani** f. sp. **pisi**; **Thielaviopsis basicola**, ID, WA.
ROT, Root; Damping-Off. **Pythium ultimum**, CA; **Rhizoctonia solani**, CA.
ROT, Stem. **Sclerotinia sclerotiorum**, ND, WA; **S. minor**, AR.
RUST. **Uromyces ciceris-arietini**, CA, GA.
VIRUS. **Mosaic**, unidentified, CA; **Chickpea filiform**, WA; **Lettuce Mosaic**, CA; **Red Clover Vein Mosaic**, WA; **Pea Enation Mosaic**, ID, WA; **Bean leaf Roll**, ID, WA; **Pea Streak**, ID, WA.
WILT. **Fusarium lateritium** f. sp. **cicerii**; Vascular Disease; **Phialophora gregata**; **Fusarium solani** f. sp. **pisi**, CA. **Verticillium albo-atrum**.

CHINESE CABBAGE (*Brassica pekinensis, B. chinensis*)
ANTHRACNOSE. **Colletotrichum higginsianum**, FL.
BACTERIAL Black Rot. **Xanthomonas campestris**, IN, MD, TX.
BACTERIAL Leaf Spot. **Pseudomonas maculicola**, VA.
BACTERIAL, MLO. **Aster Yellows**.
BLIGHT, Southern. **Sclerotium rolfsii**, TX.
CLUB ROOT. **Plasmodiophora brassicae**, CT, MA, NJ, OH, PA.

Leaf Spot, Black. **Alternaria oleracea**, CA, CT, FL, MA, NH, TX; Gray, **A. brassicae**, CA, CT, FL, IN, MD; **Cercospora brassicola**, FL, GA, NH, NJ; **Cercosporella brassicae**, white spot, widespread.
Leaf Spot; White Spot. **Pseudocercosporella capsellae**, CA.
Nematode, Root Knot. **Meliodogyne** sp.
Powdery Mildew. **Erysiphe polygoni**, AZ, MA; **Peronospora parasitica**.
Rot, Root; Damping-off. **Aphanomyces raphani**, WI.
Rot, Watery Soft. **Sclerotinia sclerotiorum**, AZ, MA, NY, TX.
Virus. **Turnip Mosaic**; **Cauliflower Mosaic**.
White Rust. **Albugo candida**, AL.

CHINESE WAXGOURD (*Benincasa*)
Anthracnose. **Colletotrichum lagenarium**, IN.
Downy Mildew. **Pseudoperonospora cubensis**, MA, OH.
Nematode, Root Knot. **Meloidogyne** sp.

CHIVES (*Allium schoenoprasum*)
Blight, Leaf. **Botrytis byssoidea**, CA.
Downy Mildew. **Peronospora destructor**, CA.
Rust. **Puccinia porri** (II, III), CT, NY, WA.
Smut. **Urocystis cepulae**, MA.

CHOKEBERRY (Aronia)
Bacterial Fire Blight. **Erwinia amylovora**, MI, TX, WV.
Dodder. **Cuscuta compacta**, FL.
Leaf Spot. **Ascochyta pirina**, WI; **Cercospora mali**, AL, TX; **C. piri**, MI, NH, WI; **Mycosphaerella arbutifolia**, NY; **Phyllosticta arbutifolia**, NJ.
Rot, Brown. **Monilinia fructicola**, WI.
Rot, Root. **Phymatotrichum omnivorum**, TX.
Rust. **Gymnosporangium clavariiforme** (0, I), MA; III on juniper; **G. clavipes** (0, I), CT, ME, MA, TX; **G. davisii** (0, I), ME, MI, NH, WI; **G. fraternum** (0, I), DE, MA, ME, NJ, PA; III on chamaecyparis.

CITRUS FRUITS (Grapefruit, Lemon, Lime, Orange)
Anthracnose; Withertip, Fruit Spot. **Glomerella cingulata**, general.
Anthracnose, Lime; Withertip. **Gloeosporium limetticolum**, on lime only, CA.
Anthracnose, Scab. **Elsinoë australis**, FL.
Bacterial Blast; Blight; Black Pit. **Pseudomonas syringae**, CA; after cold rains.
Bacterial Brown Leaf Spot. **Burkholderia andropogonis**, FL.
Bacterial Canker. **Xanthomonas campestris** pv. **citri**, FL.
Bacterial Crown Gall. **Agrobacterium tumefaciens**, AZ, CA.
Bacterial Greening. **Candidatus Liberibacter asiaticus**, Central America, Puerto Rico.
Bacterial, MLO. **Citrus Stubborn Disease**.
Blight, Brown Spot; Fruit and Young Shoots. **Alternaria citri**, FL.
Blight, Leaf and Stem. **Phytophthora syringae**, CA.
Blight, Seedling; Fruit Rot. **Sclerotium rolfsii**.

BLIGHT, Thread. **Pellicularia koleroga**, Gulf states.

BLIGHT, Twig. **Gibberella baccata** (*Fusarium lateritium*), CA; **Tryblidiella rufula**, TX.

BLOTCH, Sooty. **Gloeodes pomigena**, Gulf states.

CANKER, Branch. **Eutypella** spp., CA; **Spencermartinsia viticola, Neofusicoccum austral**, and **N. parvum**, CA.

CANKER, Branch Knot. **Sphaeropsis tumefaciens**, FL.

CANKER, Branch Wilt. **Exosporina fawcettii**, CA.

CANKER; DIEBACK. **Aspergillus foetidus**, CA.

CANKER, Felt Fungus. **Septobasidium** spp., Gulf states.

CANKER, Trunk. **Botryodiplodia theobromea**, TX.

CANKER, Wound. **Hendersonula toruloidea**, CA.

DAMPING-OFF. **Pythium** spp.; **Rhizoctonia solani**, cosmopolitan.

DODDER. **Cuscuta americana**, CA, FL; **Cassytha filiformis**, a dodderlike plant, FL.

GREENING DISEASE. **Huanglongbing**, FL.

LEAF SPOT. **Cercospora aurantia**, FL, TX; Greasy Spot, **C. citri-grisea**, FL; Tar Spot, **C. gigantea**, FL; **Cladosporium oxysporum**, FL; **Mycosphaerella lageniformis**, CA; **M. citri**, FL; **Pleospora** sp., CA; **Alternaria citri**, on Rangpur lime; Fruit Spot, **Septoria citri**, CA, TX; **S. limonium**, CA; **Xanthomonas fuscans**, Brazil.

LEAF SPOT, Algal; Red Rust. **Cephaleuros virescens**, Gulf states.

MISTLETOE. **Phoradendron** sp., Gulf states.

NEMATODE, Burrowing. **Radopholus similis**, FL.

NEMATODE, Citrus. **Tylenchus semipenetrans**, CA, FL.

NEMATODE, Lance. **Hoplolaimus coronatus**.

NEMATODE, Lesion. **Pratylenchus brachyurus**; **P. vulnus**.

NEMATODE, Ring. **Criconema civellae**; **Criconemoides citri**, FL.

NEMATODE, Stubby Root. **Paratrichodorus christiei**.

NONPARASITIC. **Blight**. Boron deficiency, CA, FL.

Blight. Boron deficiency, CA, FL.

Bronzing. Magnesium deficiency, FL.

Cancroid Spot. Genetic abnormality.

Chlorosis. Iron deficiency, AZ, CA, FL; manganese deficiency, FL.

Exanthema. Copper deficiency, augmented by excessive nitrogen fertilization and bad drainage, CA, FL.

Greasy Spot. Black Melanose, Cause unknown, CA, TX; chiefly grapefruit.

Gummosis. Gum Spot; Gummosis. Environmental injuries in part.

Leprosis. Florida Scaly Bark; Nailhead Rust. Cause unknown, FL.

Mottle Leaf. Zinc deficiency, AZ, CA, FL.

Oleocellosis. Rind-oil Spot. Chemical injury from release of oil in rind.

Rumple. Rind network on lemon; cause unknown.

Silver Scurf. Thrips injury on fruit, widespread; Silvering. Rust mite injury; Stigmonose. Insect punctures.

Wilt; Blight; Leaf Curl. Irregular water supply, FL.

PHYTOPLASMA, Citrus Huanglongbing. **Candidatus L. asiaticus**, FL.
POWDERY MILDEW. **Oidium** sp., CA, FL; **O. tingitaninum**.
ROT. **Diplodia**; Collar; Twig Blight. **Diplodia natalensis**, general.
ROT, Black, of Fruit. **Alternaria citri**, widespread.
ROT, Brown; Gummosis. **Phytophthora citrophthora**, AZ, CA, FL; **P. parasitica**, "mal di gomma," CA, FL.
ROT, Charcoal. **Macrophomina phaseoli**, AZ, CA.
ROT, Cottony Fruit; Twig Blight. **Sclerotinia sclerotiorum**, CA, FL, TX.
ROT, Dothiorella; Bark Canker. **Botryosphaeria ribis**, widespread.
ROT, Fruit. **Aspergillus** spp.; **Fusarium** spp.; **Mucor** spp.; **Oospora citri aurantii**, sour rot, **Nematospora coryli**, dry rot; **Candida krusei**; **Gloeosporium foliicolum** (*Glomerella cingulata*), FL; **Penicillium italicum**, Mexico.
ROT, Gray Mold; Twig Blight. **Botrytis cinerea**, CA.
ROT, Green Mold. **Penicillium digitatum**; Blue Mold, CA; **P. italicum**; Pink Mold, **P. roseum**, cosmopolitan; Black Mold, **Rhizopus stolonifer**; **Trichothecium viride**, occasional.
ROT. Melanose; Phomopsis; Decorticosis; Shell Bark. **Diaporthe citri**, general.
ROT, Mushroom Root. **Armillaria mellea**, CA; **Clitocybe tabescens**, FL.
ROT, Root. **Fusarium solani**; **Phymatotrichum omnivorum**, TX; **Poria vaporaria**, CA; **Thielaviopsis basicola**.
ROT, Sour. **Geotrichum citri-aurantii**, Mexico.
ROT, Wood. **Daldinia concentrica**, occasional; **Ganoderma lucidum**, FL; **Polyporus** spp.; **Schizophyllum commune**; **Trametes hydnoides**; **T. hispida**.
SOOTY MOLD, **Capnodium citri**; **C. citricola**. Gulf states.
SPOT ANTHRACNOSE; Citrus Scab. **Elsinoë fawcettii**, AL, CA, FL, LA, MS.
VIRUS. **Algerian Navel Orange**, FL; **Citrus Ringspot,** CA, FL; **Citrus Exocortis** (viroid); **Citrus Psorosis**, Scaly Bark; **Citrus Tatter Leaf**; **Citrus Tristeza**, Quick Decline; **Citrus Vein Enation**; **Citrus Xyloporosis**, Cachexia; **Citrus Yellow Vein**; **Citrus Leaf Rugose**, CA; **Citrus Tristeza**, LA; **Citrus Variegation**; **Citrus Leprosis**; **Citrus Enation**-Woody Gall, CA.

In the home garden oil sprays for scale help to get rid of sooty mold. A neutral copper spray just before growth starts is recommended in Florida for citrus scab and one just after flowering for melanose. Citrus trees in the burrowing nematode area are being pulled and the soil treated. Consult the California and Florida Agricultural Experiment Stations for the latest information on specific Citrus problems.

COLLARDS
See Cabbage.
Subject to downy mildew and Sclerotinia rot.

CORIANDER (*Coriandrum*)
ANTHRACNOSE. **Gloesporium** sp., MD.
BACTERIAL Leaf Spot. **Pseudomonas syringae**, CA, FL; **Xanthomonas campestris** pv. **carotae**, OR.
NEMATODE, Root Knot. **Meloidogyne** sp., FL.

VIRUS. **Coriander Feathery Red Vein**, CA; **Apium Virus Y**, **Potyvirus,** CA.
WILT. **Fusarium oxysporum** f. sp. **coriandrii**, CA.

CORN, SWEET (*Zea mays* var. *saccharata*)
BACTERIAL Leaf Blight. **Acidovorax avenae** syn. **Pseudomonas avenae**, IL.
BACTERIAL Leaf Stripe. **Pseudomonas andropogonis**, FL.
BACTERIAL Spot. **Pseudomonas syringae**, MA.
BACTERIAL Stalk Rot. **Erwinia dissolvens**, FL, WV.
BACTERIAL Wilt. **Erwinia stewartii**, general**; Clavibacter michiganensis** subsp. **nebraskensis**, IN, MN, TX.
BLIGHT, Leaf. **Helminthosporium turcicum** syn. **Exserohilum turcicum**), HI, IL, TX; middle Atlantic and southern states, occasional in central states; **H. maydis** (*Cochliobolus heterostrophus*), southern leaf blight; **Clavibacter michiganensis** subsp. **nebraskensis,** IN; **Pantoea agglomerans**, Mexico.
BLIGHT, Seedling; Stalk Rot; Ear Rot. **Penicillium oxalicum**.
DOWNY MILDEW. **Peronosclerospora sorghi**, IL.
LEAF SPOT. **Cercospora zeae-maydis**, VA, Ontario, Canada; **Phaeosphaeria maydis**, FL; **Physoderma maydis**, brown spot, in South; Eyespot, **Kabatiella zeae; Pantoea ananatis,** Argentina.
LEAF STRIPE. **Burkholderia gladioli**, Mexico.
LESIONS, Stem. **Rhizoctonia solani**, MS.
NEMATODE, Lesion. **Pratylenchus penetrans**.
NEMATODE, Root Knot. **Heterodera zeae**, VA; **Meloidogyne chitwoodi**, Pacific NW; **M. hapla**, Pacific NW.
ROOT DISEASE. **Fusarium oxysporum, F. solani, F. acuminatum, F. proliferatum**, MN.
ROT, Black Bundle Disease. **Cephalosporium acremonium**, MT.
ROT, Dry Ear. **Nigrospora sorghi**, IA.
ROT, Ear, Root and Stalk; Seedling Blight. **Diplodia zeae**, ME to VA, SD, TX; **Fusarium** spp.; **F. moniliforme**, cosmopolitan; **Burkholderia gladioli**, MS.
ROT, Root. **Gibberella zeae**, chiefly east of Mississippi River; **Phoma terrestris**, DE, MD; **Pythium** spp., IL, IA, OH, TX; **Helminthosporium pedicellatrum**, CA; **Exserohilum pedicellatum,** TX; **Phomus terrestris**, NC.
ROT, Stem. **Burkholderia gladioli**, Mexico.
RUST. **Puccinia sorghi** (II, III), general; 0, I on oxalis; **P. polysora**, GA, Southern U.S.
SMUT. **Ustilago maydis**, general, FL, IL, MI, MN, NY, TX,WI; **Sphacelotheca reiliana**, Head Smut.
SMUT, False. **Ustilaginoidea virens**, MS.
VIRUS. **Barley Yellow Dwarf ,** IL; **Corn Leaf Fleck**; **Corn Mosaic**, **Corn Stunt**, AL, AZ, CA, GA, IN, KY, MO, SC; **Wheat Streak Mosaic**, CO, ID, IN, KS, MI, NE, OH, TX, UT; **Cucumber Mosaic**, celery strain; **Maize Chlorotic Mottle**, HI, KS, NE; **Maize Dwarf Mosaic**, AL, AZ, AR, CA, GA, ID, IA, MO, NJ, NM, NY, OH, PA, WA; **Maize Rayado Fino**; **Maize Stripe**, FL; **Sorghum Stunt Mosaic**, CA; **Oat Blue Dwarf. Marafivirus**, North America, Great Plains.

WILT, Stunt. **Phytophthora sansomeana**, OH.
WILT, Vascular. **Pantoea agglomerans**, Mexico.
WITCHWEED. **Striga asiatica**, on roots, NC.

Smut is the most conspicuous corn disease in home gardens. Bacterial wilt is dependent on survival of insect vectors and may be serious after a warm winter unless resistant hybrids are used.

CRANBERRY (*Vaccinium,* subgenus *Oxycoccus*)
DIEBACK. **Diaporthe vaccinii**, WI.
GALL, Red Leaf. **Synchytrium vaccinii**, NJ.
GALL; Shoot Hypertrophy; Rose-bloom. **Exobasidium vaccinii**, general; **E. oxycocci**, MA.
LEAF SPOT. **Discohainesia oenotherae**, also storage rot, MA, NJ, OR, WA, WV; **Mycosphaerella nigromaculans**, black spot, OR, WA; **Phyllosticta putrefaciens**, also berry rot, MA, NJ; **Ramularia multiplex**, MI, NY, WI; **Venturia compacta**, leaf smudge, general; **Cladosporium oxycocci**, leaf mold, NJ, WA.
NEMATODE. **Atylenchus decalineatus**.
NEMATODE, Sheath. **Hemicycliophora similis**.
NEMATODE, Stubby Root. **Paratrichodorus christiei**.
POWDERY MILDEW. **Microsphaera alni** var. **vaccinii**, AL, NJ, OH.
ROT, Berry; Blotch. **Acanthorhynchus vaccinii**, ME to NC, WI, OR, WA; **Ceuthospora lunata**, black rot, MA, NJ, OR, WA, WI; **Curvularia inaequalis**, IL, NJ, WI; **Sphaeronema pomorum**, NJ; **Sporonema oxycocci**, AK, ME, MA, NJ, OR, WA, WI.
ROT, Berry Speckle. Several fungi associated.
ROT, Bitter; Leaf Spot. **Glomerella cingulata** var. **vaccinii**, general; **Phyllosticta vaccinii**, WI.
ROT, Early; Scald, Blast. **Guignardia vaccinii**, general; **Phyllosticta vaccinii**, WI.
ROT, End. **Godronia cassandrae**, general.
ROT, Fairy Ring; Root. **Psilocybe agrariella** var. **vaccinii**, MA, NJ.
ROT, Fruit. **Acanthorhyncus vaccinii** and **A. alba**, North America; **Gliocephalotrichum bulbilium**, MA, NJ.
ROT, Hard; Twig Blight. **Monilinia oxycocci**, MA, ME, OR, WA, WI.
ROT, Root; DIEBACK. **Phytophthora cinnamomi**, MA.
ROT, Storage. **Penicillium** spp., cosmopolitan; **Diaporthe vaccinii**, general; **Botrytis cinerea**, gray mold, also blossom blight; **Alternaria** sp.; **Dematium** spp.; **Melanospora destruens**, in market; **Gloeosporium minus**, MD, NJ; **Pestalotia vaccinii**; **Stemphylium ilicis**.
ROT, Witches' Broom. **Naevia oxycocci**, ME, MI, NH, NY.
RUST. **Pucciniastrum vaccinii** (II, III), OR, UT, WA; 0, I on hemlock.
VIRUS. **Cranberry False-Blossom**, ME to NJ, OR, WA, WI; **Cranberry Ring Spot,** WI.

CUCUMBER (*Cucumis sativus*)
ANTHRACNOSE. **Colletotrichem lagenarium**, general.

BACTERIAL Angular Leaf Spot. **Pseudomonas syringae** pv. **lachrymans**, general.
BACTERIAL Soft Rot. **Erwinia aroideae**; **E. carotovorus**.
BACTERIAL Spot. **Xanthomonas cucurbitae**, MA, MI.
BACTERIAL Wilt. **Erwinia tracheiphila**, general.
BLIGHT, Blossom; Fruit Rot. **Choanephora cucurbitarum**, FL, GA, NJ, OK, RI, TX.
BLIGHT, Gummy Stem; Black Fruit Rot. **Mycosphaerella citrullina**, widespread; **M. melonis**, AZ.
BLIGHT, Leaf. **Alternaria cucumerina**, general.
BLIGHT, Southern. **Sclerotium rolfsii**, VA to FL, TX.
DAMPING-OFF. **Rhizoctonia solani**, also stem rot; **Pythium** spp.
DODDER. **Cuscuta gronovii**, NY.
DOWNY MILDEW. **Pseudoperonospora cubensis**, general; U.S., Beirut, Lebanon.
LEAF SPOT. **Cercospora cucurbitae** AL, DE, IA, NJ, TX, WI; **Ascochyta** sp., OR; **Gloeosporium** sp. IL; **Phyllosticta cucurbitacearum**, DE, OH, TX; **Septoria cucurbitacearum**, DE, MA, NH, PA; **Stemphylium cucurbitacearum**, OH; **Alternaria consortiale**, WA; **Ulocladium cucurbitae**, CA, NY.
MOLD. Seed. **Alternaria tenuis**, cosmopolitan; **Curvularia trifolii**, NJ.
NEMATODE, Hop Cyst. **Heterodera humuli**, OR.
NEMATODE, Root Knot. **Meloidogyne incognita**; **M. javanica**.
NONPARASITIC. **Chlorosis**. Nitrogen, manganese, or potassium deficiency.
POWDERY MILDEW. **Golovinomyces cichoracearum**, general; **Sphaerotheca fuliginea**, CA, NC; **Leveillula taurica**, ID.
ROT, Charcoal. **Macrophomina phaseoli**, IL.
ROT, Fruit. **Fusarium** spp., TX; **Rhizopus stolonifer**, occasional. **Diplodia natalensis**; **Botrytis cinerea**.
ROT, Root. **Fusarium solani** f. sp. **radicicola**, CT, OR, WA; **Phymatotrichum omnivorum**, TX.
ROT, Stem; Fruit. **Sclerotinia sclerotiorum**, occasional.
SCAB. **Cladosporium cucumerinum**, general.
VIRUS. **Bromegrass Mosaic Virus**, general; **Cucumber Mosaic**, general; **Beet Curly Top**, CA, ID, OR, TX, UT, WA; **Tobacco Ring Spot,** MD, PA, VA; **Zucchini Yellow Mosaic**, CA, FL, NY; **Cucurbit Leaf Curl**, AZ, NM, TX.
WILT. **Fusarium oxysporum** f. sp. **cucumerinum**, FL; **F. oxysporum** f. sp. **cucurbitae**; **Verticillium albo-atrum**, CA.
WILT, Basal Stem Lesions. **Phytophthora capsici**, Mexico.

Starting cucumbers under Hotkaps and then treating to control insect vectors helps to reduce bacterial wilt and virus diseases. Choose varieties resistant to mosaic and scab.

CURRANT (*Ribes* spp.)
ANTHRACNOSE; Fruit Spot. **Pseudopeziza ribis**, general.
BLIGHT, Cane. **Bostryosphaeria ribis** var. **chromogena**, general.
BLIGHT, Gray Mold; Fruit Spot. **Botrytis cinerea**, Northeast, OR, WA.

Blight, Thread. **Pellicularia koleroga**, FL.
Canker, Cane Knot. **Thyronectria berolinensis**, CT to IN, KS, UT.
Canker; Dieback. **Nectria cinnabarina**, ME to CO, WA; **N. ditissima**, MN, NY; Black Pustule, **Phragmodothella ribesia**, Northeast, Pacific Northwest.
Downy Mildew. **Plasmopara viticola**, WI.
Leaf Spot. **Alternaria** sp., MI; **Cylindrosporium ribis**, WI; **Mycosphaerella ribis**, ME to MD, AR, OR, WA.
Leaf Spot, Angular. **Cercospora angulata**, NY to VA, KS, MN; **C. ribis**, AL, IN, IA.
Powdery Mildew. **Microsphaera grossulariae**, MT, NE, NH; **Sphaerotheca mors-uvae**, AK, CA, CT, MT, NE, OR, WA; **Phyllactinia corylea**, MI.
Rot, Berry. **Glomerella cingulata**, CT, PA.
Rot, Collar. **Fomes ribis**, NY to IN, MN, UT.
Rot, Root. **Armillaria mellea**, CA, OR, WA; **Phymatotrichum omnivorum**, TX; **Hypholoma perplexum**, NY.
Rust, White Pine Blister. **Cronartium ribicola** (II, III), ME to VA, IL, MN, OR, SD, WA.
Puccinia caricis (0, I), CT, IN, MD, NY, II, III on Carex.
Virus. **Currant Mosaic**, MD, NY.
Wilt. **Verticillum** sp., NY.

Black currant is the most important host of white pine blister rust and should not be grown; red currants should be kept at least 300 ft from pines. Cut out canes with cane blight.

CURRANT, FLOWERING (*Ribes*)
Anthracnose. **Pseudopeziza ribis**, IL, IA, MN, MT.
Bacterial Leaf Spot. **Pseudomonas ribicola**.
Blight; Dieback. **Botrytis cinerea**, AK, IN.
Canker; Dieback. **Botryosphaeria ribis**, KS; **Phragmodothella ribesia**, NY; **Thyronectria berolinensis**, KS; **Nectria cinnabarina**, AK, KS.
Leaf Spot. **Cercospora angulata**, MN; **C. ribicola**, OR, WA; **Marssonina ribicola**, CO; **Mycosphaerella aurea**, NY to KS, SD, WA; **M. ribis**, NY to KS, MN, UT; **Phyllosticta grossulariae**, IN; **Septoria san-guinea**, WA.
Rust. **Coleosporium jonesii**, MN; **Cronartium occidentale**, MT to NM, CA, WA; **C. ribicola**, ME to MD, CO, MN, NV, Pacific states; **Melampsora ribesii-purpureae**; II, III on willow; **Puccinia caricina**, NY to IA, AK, CA, OR; **P. micrantha**; **P. parkerae**.
Virus. Viruslike vein-banding, CA, CT, MN, NJ, NY, OR, RI, WA.

D

DEWBERRY (*Rubus*)

BACTERIAL Cane Gall. **Agrobacterium rubi**, NY, OR.

BACTERIAL Crown Gall. **Agrobacterium tumefaciens**, general.

BACTERIAL Hairy Root. **Agrobacterium rhizogenes**, OR.

BLACK MILDEW. **Irenina sanguinea**, AL, LA.

BLIGHT, Cane; DIEBACK. **Leptosphaeria coniothyrium**, general; **L. thomasiana**, OR, WA.

BLIGHT, Spur. **Didymella applanata**, OR, WA.

BLIGHT, Stamen; Dry Berry. **Hapalosphaeria deformans**, OR, WA.

BLOTCH, Sooty. **Gloeodes pomigena**, NC, PA.

CANKER; Cane Spot. **Ascospora ruborum**, AL, CA.

CANKER; Fruit Rot. **Glomerella cingulata**, GA, IL, MD, MS.

DOWNY MILDEW. **Peronospora rubi**, FL, MD, WI, WA; **P. potentillae**, CT, IL, LA.

FRUIT SPOT; Flyspeck. **Leptothyrium pomi**, NC.

GALL, Yellow Leaf. **Synchytrium aureum**, WI.

LEAF SPOT. **Mycosphaerella confusa**, NJ to FL, IL, TX; **Pezizella oenotherae**, MD to NC; **Phyllosticta ruborum**, NY; **P. dispergens**, IL; **Septoria rubi** (**Mycosphaerella rubi**), general; **S. darrowi**, NY.

POWDERY MILDEW. **Sphaerotheca macularis**, IL, IN, MN, OH, PA, Pacific Northwest.

ROSETTE; Double Blossom. **Cercosporella rubi**, NC to AL, LA, MS, TX, IL.

ROT, Collar. **Rhizoctonia solani**, TX, WA.

ROT, Fruit. **Botrytis cinerea**; **Phyllosticta carpogena**, MD, NJ, NY, NC.

ROT, Root. **Armillaria mellea**, OR; **Collybia dryophila**, NC; **Corticium galactinum**, MD, VA, TX; **Helicobasidium purpureum**, NC, TX; **Phymatotrichum omnivorum**, TX.

RUST. **Gymnoconia peckiana**, orange (0, I, III), ME to VA, CA, MN, MO; **Kuehneola uredinis**, yellow (0, I, II, III), ME to FL, CA, KS, TX, WA; **Kunkelia nitens**, orange (I), CA, OR, CT to FL, IA, TX; **Mainsia rubi** (II, III), TX.

SPOT ANTHRACNOSE. **Elsinoë veneta**, general.

R.K. Horst, *Field Manual of Diseases on Fruits and Vegetables*,

DOI 10.1007/978-94-007-5974-9_7,

VIRUS. **Raspberry Beta Leaf Curl**, MI, OH, TX; **Raspberry Mosaic**, CT, MI, NJ, NY, Pacific Coast; **Loganberry Dwarf**, especially on loganberry, CA, OR, WA.
WILT. **Verticillium albo-atrum**, CA, OR, WA.

Use virus-free, bacterial-free, planting stock; spray for anthracnose.

DILL (*Anethum*)
BACTERIAL, MLO. **Aster Yellows**, NY, TX.
BLIGHT. **Itersonilia perplexans**, CA.
DAMPING-OFF. **Rhizoctonia solani**, GA.
DODDER. **Cuscuta** sp., GA.
LEAF SPOT; Stem Spot. **Cercospora anethi**, ND, OR, TX; **Phoma anethi**, CT, IN, IA.
ROT, Root; Wilt. **Fusarium** sp., OH; **Phymatotrichum omnivorum**, TX.
ROT, Stem. **Sclerotinia sclerotiorum**, TX.

E

EGGPLANT (*Solanum melongena*)

Anthracnose. **Colletotrichum** spp.; **Gloeosporium melongenae**, NJ to FL, IA, TX; **Glomerella cingulata**, IA.

Bacterial Canker. **Clavibacter michiganense**, WY.

Bacterial Soft Rot. **Erwinia carotovora**, NJ.

Bacterial Wildfire. **Pseudomonas syringae** pv. **tabaci**.

Bacterial Wilt. **Pseudomonas solanacearum**, general.

Blight, Early. **Alternaria solani**, occasional, NY to FL, LA, WI.

Blight, Late. **Phytophthora infestans**, FL, NY.

Blight, Phomopsis; Fruit Rot. **Phomopsis vexans**, general.

Blight, Southern. **Sclerotium rolfsii**, VA to FL, LA, WI.

Damping-Off. **Pythium debaryanum**, CT, LA, NY; **Rhizoctonia solani**, also stem and fruit rot, general.

Dodder. **Cuscuta** spp., KS, NJ, PA, VA.

Downy Mildew. **Peronospora tabacina**, SC.

Fruit Spot. **Diplodia natalensis**, FL; **Lasiodiplodia theobromae**, GA.

Leaf Spot. **Ascochyta lycopersici**, DE, IL, IN, NY; **Cercospora melongenae**, CA; **Phyllosticta solani**, LA; **P. hortorum**, LA, NJ; **Septoria lycopersici**, IN, MD, NC, VA; **Stemphylium solani**, FL.

Nematode, Golden. **Globodera rostochiensis** (formerly, **Heterodera rostochiensis**).

Nematode, Lesion. **Pratylenchus pratensis**, TX.

Nematode, Root Knot. **Meloidogyne hapla**; **M. incognita**; **M. javanica**.

Nematode, Tobacco Cyst. **Heterodera tabacum**, VA.

Powdery Mildew. **Golovinomyces cichoracearum**, NJ, VA.

Rot, Charcoal. **Macrophomina phaseoli**, NJ.

Rot, Collar. **Fusarium solani** f. sp. **eumartii**, CA.

Rot, Cottony Leak. **Pythium aphanidermatum**, CA, FL, TN.

Rot, Fruit. **Colletotrichum truncatum**, MS; **Phytophthora parasitica**, FL, IN; **Rhizopus stolonifer**, CA, IN, TX.

Rot, Gray Mold. **Botrytis cinerea**, CA, CT, MA, NJ, VA, WA.

Rot, Root. **Phymatotrichum omnivorum**, TX.

R.K. Horst, *Field Manual of Diseases on Fruits and Vegetables*,
DOI 10.1007/978-94-007-5974-9_8, © Springer Science+Business Media Dordrecht 2013

Rust. **Puccinia substriata (I)**, AL, FL, GA, IA, TX.
Virus. **Eggplant Mosaic**; **Cucumber Mosaic**; **Beet Curly Top**, OR, TX, WA, **Tomato Spotted Wilt**.
Wilt. **Fusarium** sp.; **Verticillium albo-atrum**, general.

Phomopsis blight and Verticillium wilt are the two most important eggplant diseases. Choose varieties resistant to blight and for wilt plant a long rotation, which does not include tomatoes, potatoes, or raspberries.

ELDERBERRY
Virus. **Tobacco Ring Spot**.

ENDIVE, ESCAROLE, WITLOOF CHICORY (*Cichorium*)
Bacterial, Apical Rot. **Pseudomonas cichorii**, CA.
Bacterial Center Rot. **Pseudomonas cichorii** and **P. intybus**, AZ, CA, FL, MT, TX, WA.
Bacterial, Leaf Spot. **Pseudomonas cichorii**, FL; **P. syringae**, CA.
Bacterial, MLO. **Aster Yellows** and **California Aster Yellows**.
Bacterial Soft Rot. **Erwinia carotovora**, MA, NY.
Blight, Southern. **Sclerotium rolfsii**, FL, TX.
Damping-Off; Bottom Rot, Leaf Blight. **Rhizoctonia solani**, CA, CT, FL, NY, TX.
Downy Mildew. **Bremia lactucae**, FL, PA.
Leaf Spot. **Alternaria** sp., CA, CT, FL, NY; **Cercospora cichorii**, TX; **Marssonina panattoniana**, TX; **Ramularia cichorii**, NY.
Nematode, Root Knot. **Meloidogyne** sp., MA, NJ, TN.
Nematode, Sting. **Belonolaimus gracilis**.
Nonparasitic. **Brown Heart**. Boron deficiency, in part, NJ, NY.
Tipburn. High temperature and excessive transpiration.
Powdery Mildew. **Golovinomyces cichoracearum**, ID, NJ, CA.
Rot, Crown Wilt. **Sclerotinia minor**, CA.
Rot, Gray Mold. **Botrytis cinerea**, CA, FL, NY, PA.
Rot, Root. **Pythium debaryanum**, CT, FL, NY, PA; **Phymatotrichum omnivorum**, AZ.
Rot, Watery Soft. **Sclerotinia sclerotiorum**, AZ, AR, CA, FL, LA, MT, PA, TX.
Rust. **Puccinia hieracii** (0, I, II, III), CA, CT, MA, NY.
Slime Mold. **Fuligo septica**, NJ.
Virus. **Mosaic**, unidentified, FL; **Tomato Spotted Wilt**; **Tomato Ringspot,** VT; **Tobacco Streak**, FL.

F

FENNEL (*Foeniculum*)

BACTERIAL Leaf Spot. **Pseudomonas syringae**, CA.

BACTERIAL Soft Rot. **Erwinia carotovora**, IL.

BLIGHT, Stem, Foliage. **Cercosporidium punctum**, CA.

DAMPING-OFF; Stem Pitting. **Rhizoctonia solani**, GA, NJ.

NEMATODE, Root Knot. **Meloidogyne** sp., FL.

ROT, Stem. **Sclerotinia sclerotiorum**, IL, NJ, TX; **Sclerotinia minor**, CA.

FIG (*Ficus carica*)

ANTHRACNOSE; Fruit Rot. **Colletotrichum gloeosporioides** (*Glomerella cingulata*), NC to TX.

BACTERIAL Crown Gall. **Agrobacterium tumefaciens**, CA, FL, TX.

BLOTCH, Leaf. **Cercospora fici**, NC to FL, TX.

BLIGHT, Limb. **Corticium salmonicolor**, Gulf states.

BLIGHT, Southern. **Sclerotium rolfsii**, FL.

BLIGHT, Thread. **Pellicularia koleroga**, FL, LA, MS; Web, **P. microsclerotia**, FL to LA, TX.

BLIGHT, Twig. **Gibberella baccata**, CA.

CANKER; DIEBACK. **Botryosphaeria ribis**; **Diplodia sycina**, NC, OR; **Macrophoma fici**, also fruit dry rot, LA, NC, TX; **Nectria cinnabarina**, TX; **Megalonectria pseudotrichia**, LA, TX; **Physalospora rhodina**, AL, FL, TX; **Sclerotinia sclerotiorum**, CA, TX.

CANKER, Felt Fungus. **Septobasidium** spp., MS.

CANKER, Phomopsis. **Phomopsis cinerescens**, CA.

LEAF SPOT. **Colletotrichum elastica**, FL; **Corynespora cassiicola**, FL; **Ascochyta caricae**, CA; **Alternaria** sp., SC; **Cephalosporium acremonium**, LA; **Eutypa** sp., TX; **Mycosphaerella bolleana**, NC to TX; **Ormathodium fici**, LA.

NEMATODE, Dagger. **Xiphinema index**, CA.

NEMATODE, Fig Cyst. **Heterodera fici**, CA.

NEMATODE, Lesion. **Pratylenchus vulnus**, **P. musicola** and **P. pratensis**, CA.

NEMATODE, Pin. **Paratylenchus hamatus**, CA.

R.K. Horst, *Field Manual of Diseases on Fruits and Vegetables*,
DOI 10.1007/978-94-007-5974-9_9,

NEMATODE, Root Knot. **Meloidogyne** spp., southern states to CA.
NONPARASITIC. **Chlorosis**. Manganese deficiency, FL.
Little Leaf. Zinc deficiency, CA.
Sunburn. Low temperature injury to trunk and branches, CA.
ROT, Fruit. **Alternaria tenuis**, in market; **Aspergillus niger**, black mold; **Botrytis cinerea**; **Choanephora cucurbitarum**, TX; **Cladosporium herbarum**, market; **Diplodia natalensis**, TX; **Fusarium moniliforme**, CA; **Oospora** sp., sour rot, TX; **Rhizopus nigricans**, CA, Gulf states; **Trichothecium roseum**, Gulf states.
ROT, Root. **Armillaria mellea**, CA; **Dematophora necatrix**, NC; **Phymatotrichum omnivorum**, TX; **Rhizoctonia** sp., TX.
RUST. **Cerotelium** (*Physopella*) **fici** (II, III), NC to FL, AR, TX.
SOOTY MOLD. **Capnodium** sp., TX; **Fumago vagans**, Gulf states.
VIRUS. **Fig Mosaic**, AZ, CA, GA, TX, VA.
WILT, Branch. **Hendersonula toruloides**, CA.

FILBERT
See Hazelnut.

G

GARLIC (*Allium sativum*)
See Onion.

Rot, White. **Sclerotium cepivorum**, ME.
Rust. **Puccinia allii**, OR; on wild garlic, CA.
Virus. **Onion Yellow Dwarf, Leak Yellow Stripe, Garlic Common Latent**, OR, WA, Argentina; **Leaf Yellow Strip, Mosaic, Dwarfism**, Guanajuato, Mexico; **Leaf Yellow Strip Virus, Iris Yellow Spot** and **Irish Yellow Spot Virus,** OR, WA; **Japanese Garlic Virus** (Virus B and D), Pacific Northwest; **Shallot Latent**, Brazil.

Subject to diseases of onion. White rot is serious in Louisiana; pink root is a common problem. Clove rot is caused by *Penicillium digitatum*.

GINGER, EDIBLE (*Zingiber*)
Bacterial, Rhizome Rot. **Enterobacter cloacae**, HI.

GINSENG (*Panax*)
Blight. **Alternaria panax**, general; **Botrytis cinerea**, NY to NC, MI, WA.
Damping-Off. **Pythium debaryanum**, NY; **Rhizoctonia solani**, AR, IN, MI, NJ, NY, WA; **Phytophthora citricola**, MI.
Leaf Spot. **Septoria** sp.; **S. araliae**, WI; **Colletotrichum dematium**, secondary, NY to NC, MN, MO.
Nematode, Root Knot. **Meloidogyne** sp., CT, MI, NY, OH, PA, WI.
Nonparasitic. **Papery Leaf**. Moisture deficiency, sunscald, MI, MO, NJ, NY, PA.
Powdery Mildew. **Erysiphe panax**.
Rot, Rhizome. **Sclerotinia sclerotiorum**, white rot, MI, NY, OH, PA, WA, WI; **S. smilacina**, black rot, MI, MN, NY, WI.
Rot, Root. **Armillaria mellea**, WA; **Fusarium scirpi**, NY to AL, MO, WA, WI; **Phytophthora colocasiae**, NC; **Thielaviopsis basicola**, black root, IL, MI, NJ, NY, OH; **Ramularia** spp., MI, NY, OR, WA, WI.
Rot, Stem, Root; Downy Mildew. **Phytophthora cactorum**, CT to NC, IA, MI, WA.
Rust. **Puccinia araliae** (III), MA, PA.
Wilt. **Verticillium albo-atrum**, IN, KY, MI, NJ, NY, OH, PA, TN, WI.

R.K. Horst, *Field Manual of Diseases on Fruits and Vegetables*,
DOI 10.1007/978-94-007-5974-9_10, © Springer Science+Business Media Dordrecht 2013

GOOSEBERRY (*Ribes*)

ANTHRACNOSE. **Pseudopeziza ribis**, general.

BLIGHT, Cane. **Botryosphaeria ribis**, NJ, VA; **Leptosphaeria coniothyrium**, IN, MO.

BLIGHT, Thread. **Pellicularia koleroga**, FL.

CANKER; DIEBACK. **Nectria cinnabarina**, coral spot; **Botrytis cinerea**, CT, OR, WA; **Phragmodothella ribesia**, twig knot; **Physalospora obtusa**, NY to VA, KS.

DODDER, **Cuscuta** sp., MN, NY.

DOWNY MILDEW. **Plasmopara ribicola**, MT, OR, WV, WI.

LEAF SPOT. **Marssonina grossulariae**, OH, WI; **Cercospora angulata**, NY to MI, MO; **Mycosphaerella ribis**, MA to VA, AL, KS, MN, OR; **Phyllosticta grossulariae**, CT, NJ, WA, WI; **Ramularia** sp., MI.

NEMATODE, Leaf, Bud. **Aphelenchoides ritzemabosi**, CA.

NONPARASITIC. **Leaf Blotch**. Magnesium deficiency, OR.

Leaf Scorch. Potassium deficiency.

POWDERY MILDEW. **Sphaerotheca macularis**, MN; **S. mors-uvae**, general; **Microsphaera grossulariae**, CA; **Phyllactinia corylea**, NY.

ROT, Root. **Armillaria mellea**, OR, WA; **Dematophora** sp.; **Xylaria hypoxylon**, OR; **Phymatotrichum omnivorum**, TX.

RUST. **Puccinia caricina** var. **grossulariata** (0, I) leaves, fruit, ME to MD, AK, ID, KS, MS, II, III on *Carex*; **P. caricina** var. **uniporula**, IA, MD, NY, WI; **Coleosporium jonesii** (II, III), CO, MO, MN, NM, WI, WY; **Cronartium occidentale**, AZ, CO, UT; **C. ribicola**, white pine blister rust (II, III) – European varieties are resistant.

SPOT ANTHRACNOSE, Scab. **Sphaceloma ribis**, WA.

VIRUS. **Mosaic**. Unidentified, IL, NY.

Powdery mildew caused by *Sphaerotheca mors-uvae* is probably the most important gooseberry disease but is readily controlled by a lime-sulfur spray immediately after bloom, followed by bordeaux mixture spray for leaf spots.

GRAPE (*Vitis*)

ANTRACNOSE. **Elsinoë ampelina**, MI.

BACTERIAL Crown Gall. **Agrobacterium tumefaciens**, widespread.

BACTERIAL Leaf Spot. **Xanthomonas** sp., NY.

BACTERIAL, Melting, Decay. **Bacillus subtilis**, CA; **Cryptococcus laurentii**, CA.

BACTERIAL, Xylem-limited. **Pierce's Disease**, NC.

BLACK FOOT. **Cylindrocarpon liriodendra**, CA.

BLACK GOO. **Phaeoacremonium chlamydosporum**, CA.

BLACK LEG. **Cylindrocarpon obtusisporum**, CA.

BLIGHT, Shoot. **Sclerotinia sclerotiorum**, CA.

BLOTCH, Leaf. **Briosia amphelophaga**, TX.

CANKER. **Lasiodiplodia crassispora**, CA; **Diplodia corticola**, CA.

CANKER, Bot. **Lasiodiplodia theobromae**, **Diplodia seriata**, CA, Mexico.

CANKER; Deadarm; Branch Necrosis. **Cryptosporella** (*Phomopsis*) **viticola**, widespread; **Aspergillus niger**, CA.

CANKER; DIEBACK. **Botryosphaeria rhodina** and **B. obtusa**, Baja, Mexico.
CANKER, Wedge-shaped. **Botryosphaeria** sp., CA.
DECLINE. **Phaeoacremonium aleophilum**; **P. chlamydosporum**, NY, PA; **P. inflatipes**, CA; **Botryosphaeria liberica, B. viticola**, CA.
DIEBACK. **Eutypa armeniacae**, CA, NY; **E. lata**, CA, MI; **Eutypella vitis**, MI.
DIPLODIA CANE DIEBACK. **Lasiodiplodia theobromae**, Bolivia.
DOWNY MILDEW. **Plasmopara viticola**, general, serious in East.
FRUIT SPOT; Flyspeck. **Leptothyrium pomi**, PA, WV.
LEAF SPOT. **Mycosphaerella personata** (*Isariopsis clavispora*) wide-spread **Phomopsis viticola**, NY; **Septoria ampelina**, NY, TX, VA; **Septosporium heterosporum**, CA.
LEAF SPOT, Zonate. **Cristulariella pyramidalis**, WV; **C. moricola**, GA.
MEASLES. **Togninia fraxinopennsylvanica** Anamorph, **Phaeoacremonium viticola**, CA.
NEMATODE. **Meloidoderita** sp., NY.
NEMATODE, Citrus. **Tylenchulus semipenetrans**, CA.
NEMATODE, Cyst. **Heterodera punctata**, MI.
NEMATODE, Dagger. **Xiphinema index**.
NEMATODE, Lesion. **Pratylenchus pratensis**, CA; **P. minyus**, CA; **P. coffeae**, CA; **P. vulnus**, CA.
NEMATODE, Pin. **Paratylenchus hamatus**.
NEMATODE, Ring. **Criconemoides xenoplax**, CA.
NEMATODE, Root Knot. **Meloidogyne** spp.
NEMATODE, Spiral. **Helicotylenchus pseudorobustus**, CA.
NEMATODE, Stubby Root. **Paratrichodorus christiei**.
NONPARASITIC. **Little Leaf**. Zinc deficiency, CA.
Shot Berry. Defective pollination.
Skin Blanching. Sulfur dioxide injury.
PHYTOPLASMA, Aster Yellows. **Candidatus Phytoplasma asteris**, British Columbia, Ontario, Quebec, Canada.
PHYTOPLASMA, Bois Noir. British Columbia, Canada.
PIERCE'S DISEASE. **Xyella fastidiosa**, OK.
POWDERY MILDEW. **Uncinula necator**, general, MA.
ROT, Bitter. **Greeneria uvicola**, MS; **Melanconium fuligineum**, wide-spread, often secondary after black rot.
ROT, Black. **Guignardia bidwellii**, general.
ROT, Charcoal. **Macrophomina phaseoli**, TX.
ROT, Fruit. **Alternaria** sp., CA; **Aspergillus niger**, black mold, CA, OR; **Botrytis cinerea**, gray mold, CA; **Cladosporium** sp., green mold; **Penicillium** spp., blue mold, cosmopolitan; **Glomerella cingulata**; **Colletotrichum acutatum**; **C. gloeosporioides**, ripe rot; **Pestalotia** sp., **Phoma** spp.; **Botryosphaeria dothidea**, **Macrophoma** rot, MS; **Monilinia fructicola**, brown rot, British Colombia, Canada.
ROT, Root. **Armillaria mellea**, AR, CA, MO, TX, WA; **Phymatotrichum omnivorum**, TX; **Clitocybe tabescens**, SC to TX, OK; **Roesleria hypogaea**, NY to VA,

IA, MO; **R. subterranean**, MI, Midwest U.S.; **Rosellinia necatrix**, AL, IN, MI, NY, OH.

ROT, Sour. **Hanseniaspora uvarum**, Ontario Canada; **Gluconobacter cerinus**, Ontario, Canada; **Aspergillus carbonarius**, CA.

ROT, Summer Bunch. **Diplodia viticola** (*D. natalensis*).

ROT, White; Dieback. **Coniothyrium diplodiella**, MA to FL, TX.

ROT, Wood. **Schizophyllum commune**, VA; **Stereum** spp.; **Poria** spp.; **Polyporus** spp.; **Pleurotus ostreatus**, CA.

RUST. **Physopella ampelopsidis** (*P. vitis*) (II), FL, SC.

SCORCH, Leaf. **Pseudopezicula tetraspora**, NY, PA.

SPOT ANTHRACNOSE; Bird's Eye Rot. **Elsinoë ampelina**, widespread.

TRUNK DISEASE. **Neofusicoccum parvum**, **Cryptovalsa ampelina** and **Eutypella leprosa**, Chile.

VASCULAR; FOLIAR DISEASE. **Phaeoacremonium**, CA.

VIRUS. **Alfalfa Dwarf**; **Grape Fanleaf,** WA, Pacific Northwest; **Grape Leafroll** (White Emper or Disease), CA, ID, WA, Chile; **Grape Yellow Mosaic**; **Virus Complex** (Kober Stem Grooving), CA; **Tobacco Ring Spot**, MI, NY; **Tomato Ring Spot**, NY; **Grapevine Fleck**, WA; **Grapevine Stem Pitting** Associated, CA; **Peach Rosette Mosaic**; **Tomato Spotted Wilt**; **Grapevine Rupestris Vein Feathering**, CA; **Grapevine Syrah**, CA, WA, Chile.

Black rot is the most destructive grape disease in most sections of the country, often causing total loss of fruit in home gardens. Pierce's disease has destroyed many vineyards in California, the pathogen is transmitted by grafting and by leafhoppers.

GRAPEFRUIT
See Citrus Diseases.

H

HAZELNUT, FILBERT (*Corylus*)

BACTERIAL Blight. **Xanthomonas corylina**, OR, WA; **Pseudomonas syringae** pv. **coryli**.

BACTERIAL Crown Gall. **Agrobacterium tumefaciens**, WA.

BACTERIAL Spot. **Pseudomonas colurnae**, IL.

BLIGHT, Eastern Filbert. **Anisogramma anomala**, NJ, NY, OR, WA.

CANKER; Twig Blight. **Apioporthe anomala**, eastern and central states; **Hymenochaete agglutinans**, ME; **Gleosporium** sp.

FILBERT STUNT. Undetermined causal agent, OR.

KERNEL SPOT, Yeast Spot. **Nematospora coryli**, OR.

LEAF BLISTER. **Taphrina coryli**, CT, MA, WI.

LEAF SPOT. **Cercospora corylina**, OK; **Cylindrosporium vermiforme**, WI; **Gloeosporium coryli**, ME to NJ, OR, PA, WA; **Phyllosticta coryli**, widespread; **Septogloeum profusum**, IN, MA, MS; **Septoria corylina**, MA to NJ, OR, WA, WI; **Sphaerognomonia carpinea**, GA; **Gnomoniella coryli**, general; **Alternaria arborescens**, **A. alternata**, **A. tenuissima** (leaf spot and brown/gray apical necrosis), AZ.

NONPARASITIC. **Bitter Pit** and **Brown Stain** of nuts. Cause unknown, OR, WA. **Shrivel**, "Blanks." Defective pollination, OR, WA.

POWDERY MILDEW. **Phyllactinia corylea**, northeastern and north central states, OR, WA; **Microsphaera alni**, northeastern and north central states; **P. gutata**, WA.

ROT, Root. **Armillaria mellea**, CA, OR, WA; **Phymatotrichum omnivorum**, TX.

ROT, Wood. **Poria ferrea**, CA.

VIRUS. **Apple Mosaic**, OR; **Clover Yellow Edge Phytoplasma**, OR.

Bacterial blight often kills young trees in Washington and Oregon. Powdery mildew is important in the Northwest.

HORSE-RADISH (*Armoracia*)

BACTERIAL Crown Gall. **Agrobacterium tumefaciens**, NJ, NY.

BACTERIAL Leaf Spot. **Xanthomonas campestris** pv. **armoraciae**, CT, IL, IA, MD, MO, SD, VA; **X. phaseoli**, IL.

R.K. Horst, *Field Manual of Diseases on Fruits and Vegetables*,
DOI 10.1007/978-94-007-5974-9_11,

Bacterial, MLO. Root, Brittle, disease, virescence. **Spiroplasma citri**, CA, IL, MD.
Bacterial Soft Rot. **Erwinia carotovora**, in stored roots, CT, NY.
Club Root. **Plasmodiophora brassicae**, IL.
Downy Mildew. **Peronospora parasitica**, AL, IL, NJ, SD, WI.
Leaf Spot. **Alternaria brassicae**, CT to DE, IL, IA; **A. oleracea**, CT to DE, MO, NE, TX; **Cercospora armoraciae**, general; **Phyllosticta decidua**, TX, WA, WI; **P. orbicula**, NY; **Ramularia armoraciae**, general.
Nematode, Root Knot. **Meloidogyne** sp., IL, MS, OK.
Powdery Mildew. **Erysiphe polygoni**, CA.
Rot, Blue Mold. **Penicillium hirsutum**, IL, NJ.
Rot, Collar, Root. **Pellicularia filamentosa**, IL, MI, NJ, NY, TX, WA.
Rot, Root. **Phymatotrichum omnivorum**, TX; **Thielaviopsis basicola**, KS, NJ.
Rot, Root Complex. **Fusarium acuminatum**, **Verticillium dahliae** and **Pseudomonas fluorescens**, IL, WI.
Virus. **Beet Curly Top**, CA, IL, KS, OR, UT, WA; **Turnip Mosaic**, widespread.
White Rust. **Albugo candida**, probably general.
Wilt. **Verticillium albo-atrum**, MI, WA.

HUCKLEBERRY (*Vaccinium*)
Blight, Sudden Oak Death. **Phytophthora ramorum**, CA.
Virus. **Blueberry Scorch**. British Columbia, Canada.

K

KIWI (*Actinidia*)
BACTERIAL Canker. **Pseudomonas syringae**, CA.
NEMATODE, Root Knot. **Meloidogyne incognita**, SC.
ROT, Root and Wilt. **Cylindrocladium crotalariae**, SC; **Pythium ultimum**, AL.
WILT, **Verticillium albo-atrum**, Chile.

R.K. Horst, *Field Manual of Diseases on Fruits and Vegetables*,
DOI 10.1007/978-94-007-5974-9_12,

L

LEEK
See Shallot.

LEMON
See Citrus Fruits.

LENTIL (*Lens*)
ANTHRACNOSE. **Colletotrichum truncatum**, ND.
BACTERIAL, Pink Seed. **Erwinia rhapontici**, Saskatchewan, Canada.
NEMATODE, Lesion. **Pratylenchus neglectus, P. thornei**, ID.
POWDERY MILDEW. **Erysiphe trifolii**, WA.
ROT, Root. **Thielaviopsis basicola**, ID, WA; **Pythium irregulare**, WA; **Aphanomyces euteiches**, ID.
VIRUS. **Red Clover Vein Mosaic**, OR; **Virus epidemic**, WA.

LETTUCE (*Lactuca*)
ANTHRACNOSE. **Marssonina panattoniana**, NY to FL, CA, MI, OR, TX, WA.
BACTERIAL Leaf Spot. **Xanthomonas campestris** pv. **vitians**, OH.
BACTERIAL Marginal Leaf Blight. **Pseudomonas fluorescens** pv. **marginalis**, KS, MO, NJ, NY; **P. cichorii**, NY.
BACTERIAL, MLO. **Aster Yellows** (white heart, Rio Grande Disease) and **California Aster Yellows**, widespread.
BACTERIAL Rot. **Pseudomonas viridilivida**, DE, LA, MI, NH, NY, VA, WA.
BACTERIAL Soft Rot. **Erwinia carotovora**, cosmopolitan in market.
BACTERIAL Wilt. **Xanthomonas vitians**, NJ, NM, NY, PA, SC, VA.
BLIGHT, Southern. **Sclerotium rolfsii**, CA, FL, NC, SC, TX, VA.
CORKY ROOT. **Rhizomonas suberifaciens**, FL.
DAMPING-OFF; Stump Wilt; Stunt. **Pythium** spp., cosmopolitan.
DOWNY MILDEW. **Bremia lactucae**, general.
LEAF SPOT. **Alternaria** sp., secondary; **Cercospora longissima**, FL, IL, IN, TX, VA, WI; **Septoria lactucae**, occasional in East and central states, to CO, FL, MN.
LEAF YELLOWING; STUNT. **Pythium uncinulatum**, CA.

R.K. Horst, *Field Manual of Diseases on Fruits and Vegetables*,
DOI 10.1007/978-94-007-5974-9_13, © Springer Science+Business Media Dordrecht 2013

NECROSIS OF ROOTS, SYSTEMIC. **Plasmopara lactucaeradicis**, hydroponically grown, VA.
MOLD, Leaf; Seed. **Pleospora herbarum**, FL, KY, NY.
NEMATODE. **Naccobus batatiformis**.
NEMATODE, Lesion. **Pratylenchus** sp.
NEMATODE, Root Knot. **Meloidogyne hapla**.
NEMATODE, Sting. **Belonolaimus gracilis**.
NONPARASITIC. **Tipburn**. High temperature and excessive transpiration.
Brown Blight. Cause unknown, AZ, CA.
POWDERY MILDEW. **Golovinomyces cichoracearum**, CA, MI.
ROT, Basal. **Phoma exigua**, CA.
ROT, Black Root. **Thielaviopsis basicola**, CA.
ROT, Bottom; Damping-off. **Rhizoctonia solani**, general.
ROT, Gray Mold. **Botrytis cinerea**, chiefly in greenhouses, sometimes outdoors.
ROT, Root. **Fusarium** sp., KY, OH; **Phymatotrichum omnivorum**, AZ, TX; **Phytophthora cryptogea**, CA; **P. drechsleri**, Mexico; **Pythium myriotylum**, VI.
ROT, Watery Soft; Drop. **Sclerotinia sclerotiorum**, widespread; **S. minor**, in transit and market.
RUST. **Puccinia dioicae** (0, I), IN, MN, ND, WI; II, III on *Carex*; **P. hieracii** (II), CA.
SLIME MOLD. **Physarum cinereum**, occasional under glass.
VIRUS. **Cucumber Mosaic**, NY; **Sonchus Yellow Net**, FL; **Tobacco Rattle**, CA; **Lettuce Mosaic**, general; **Radish Yellows**; **Lettuce Big Vein**, associated with *Olpidium brassicae* in roots, FL; **Tomato Spotted Wilt**, CA, HI, TX; **Tobacco Necrosis**; **Lettuce Infectious Yellows**, AZ, CA; **Lettuce Speckles Mottle**, CA; **Tomato Bushy Stunt**, CA; **Tobacco Streak**, FL; **Potato Leafroll**. Pacific Northwest, **Cucurbit Yellow Stunting Disorder**, CA; **Impatiens Necrotic Spot**, CA.
WHITE RUST. **Albugo** sp., TX.
WILT, Leaf Blight. **Pythium tracheiphilum**, WI; **Fusarium oxysporum** f. sp. **lactucum**, AZ, CA.

Nonparasitic tipburn is the most general of lettuce diseases, prevalent at high temperatures when soil is deficient in moisture. Some hot weather varieties are rather resistant. Spacing plants well apart in a well-drained soil will reduce bottom rot and drop. Eliminate wild weed hosts and spray for leafhopper vectors to reduce aster yellows.

LOGANBERRY
See Blackberry.

M

MACADAMIA

CANKER. **Phytophthora cinnamomi**, CA.
DECLINE, Quick. **Phytophthora tropicalis**, HI.

MANGO (*Mangifera*)

ANTHRACNOSE; Flower and Twig Blight; Fruit Rot. **Glomerella cingulata**, cosmopolitan.
BLIGHT, Twig. **Phomopsis** sp., FL; **Physalospora** (*Diplodia*) spp.
CANKER, Brown Felt. **Septobasidium pilosum** and **S. pseudopedicellatum**, FL.
DIEBACK. **Neofusicoccum parvum**, Peru.
GALL, ON TREE. **Fusarium decemcellulare**, FL.
LEAF SPOT. **Phyllosticta mortoni**, FL, TX; **Septoria** sp., FL.
LEAF SPOT, Algal, Green Scurf. **Cephaleuros virescens**, general.
NONPARASITIC. **Little Leaf**. Zinc deficiency, FL.
Soft Nose. On Indian varieties.
POWDERY MILDEW. **Oidium mangiferae**, CA, FL.
ROT, Root. **Phymatotrichum omnivorum**, TX.
ROT, Root and Crown. **Phytophthora palmivora**, AZ.
ROT, Stem-end Fruit. **Diplodia** sp.
SOOTY MOLD. **Capnodium** spp., general.
SPOT ANTHRACNOSE; Mango Scab. **Elsinoë mangiferae**, FL.
WILT. **Verticillium albo-atrum**, FL.

MELON, MUSKMELON, CANTALOUPE, CASSABRA (*Cucumis melo*)

ANTHRACNOSE. **Colletotrichum lagenarium**, general in East and South to AZ, CO, ND; **Marssonina melonis**, NY.
BACTERIAL Angular Leaf Spot. **Pseudomonas lachrymans**, CA, DE, IA, MD, MI, NJ, PA; **P. pseudoalcaligenes** subsp. **citrulli**, GA.
BACTERIAL Fruit Blotch. **Acidovorax avenae** (syn. **Pseudomonas pseudoalcaligenes** subsp. **citrulli**), TX.
BACTERIAL Rind Necrosis. **Erwinia** sp., GA, TX.
BACTERIAL Soft Rot. **Erwinia aroideae** and **E. carotovora**.

R.K. Horst, *Field Manual of Diseases on Fruits and Vegetables*,
DOI 10.1007/978-94-007-5974-9_14,

BACTERIAL Wilt. **Erwinia tracheiphila**, general east of Rocky Mts.; AZ to ID, WA.
BLIGHT, Gummy Stem; Black Rot. **Mycosphaerella citrullina** (*M. melonis*), DE, FL, MA, NJ, NY, TX.
BLIGHT, Leaf; Black Mold. **Alternaria cucumerina**, general.
BLIGHT, Southern. **Sclerotium rolfsii**, VA and OH to FL, TX.
BLIGHT, Stem Gumming and Stem-End Rot. **Diplodia natalensis**, TX.
DAMPING-OFF. **Rhizoctonia solani**, CA, GA, NJ; fruit rot, FL, TX; **Pythium debaryanum**, CA, CT, IA, NJ, NY; **Fusarium equiseti**, CA.
DECAY, Postharvest. **Epicoccum nigrum**, OK.
DODDER. **Cuscuta arvensis**, MD.
DOWNY MILDEW. **Pseudoperonospora cubensis**, general.
LEAF SPOT. **Cercospora** sp., CO, GA, TX; **Phyllosticta** sp., GA, OH, TX; **Septoria cucurbitacearum**, DE, MA, MI, NH, NY, PA, VT, WI; **Alternaria alternate** f. sp. **cucurbitae**, MD, Mid-Atlantic, U.S.
LEAF SPOT, STEM NECROSIS, MELON NECROTIC SPOT VIRUS, Mexico, Guatemala, Honduras, Panama, and Uraguay.
NEMATODE, Root Knot. **Meloidogyne arenaria**, **M. hapla**, NJ to FL, CA.
NONPARASITIC. **Leaf Spot**. Magnesium deficiency, NY.
POWDERY MILDEW. **Golovinomyces cichoracearum**, general.
ROT, Charcoal. **Macrophomina phaseoli**, CA, OR, TX.
ROT, Corky Dry. **Fusarium semitectum**, FL.
ROT, Fruit. **Fusarium** spp., general in market, also **Alternaria** spp.; **Monilia sitophila**, IN, NY; **Mucor** sp., NY; **Penicillium** spp., blue mold; **Phytophthora** spp.; **Rhizopus stolonifer**, cosmopolitan; **Trichoderma viride**, green mold; **Trichothecium roseum**, pink mold, occasional in market.
ROT, Root. **Monosporascus cannonballus** and **Acremonium** sp., CA; **Phymatotrichum omnivorum**, TX; **Pythium periplocum**, CA.
ROT, Root; Cottony Leak. **Pythium aphanidermatum**, AZ, CA, TX.
ROT, Stem. **Sclerotinia sclerotiorum**, AR, MA, TX; **Cephalosporium** sp.
SCAB. **Cladosporium cucumerinum**, occasional, East and central states.
VIRUS. **Beet Curly Top**, AZ, CA, ID, OR, TX, WA; **Cucumber Mosaic**, general; **Muskmelon Mosaic**, general; **Tobacco Ring Spot,** KS, MD, NC, PA, WI; **Squash Mosaic**; **Watermelon Mosaic**; **Zucchini Yellow Mosaic**, CA, FL, NY; **Lettuce Infectious Yellows**, AZ, CA, TX; **Muskmelon Vein Necrosis**, CA; **Melon Necrotic Spot**; **Melon Leaf Curl**, CA; **Papaya Ring Spot,** LA; **Watermelon Mosaic**, LA; **Cucurbit Leaf Curl**, AZ, TX; **Cucurbit Yellow Stunting**, AZ, TX, Mexico; **Melon Severe Mosaic**, Mexico, State of Guerrero; **Potyvirus**, CA.
WILT. **Fusarium oxysporum** f. sp. **melonis**, general Race 1 Colima, Mexico; **Verticillium alboatrum**, CA, OR.

Melons belong to the cucurbit family and in general have the same diseases as cucumbers. Downy mildew is a problem on the moist East Coast, and powdery mildew may be a limiting factor in the arid Southwest. Although there are varieties resistant to powdery mildew, different physiological races of the fungus keep things complicated. Sulfur dust should not be used except on sulfur-resistant melon varieties. Mosaic is transmitted both by seed and aphids; purchase virus-free seed, eliminate weeds, and keep down insects.

MONTIA (Indian Lettuce)
SMUT, Seed. **Ustilago claytoniae**, WA.

MULBERRY (*Morus*)
BACTERIAL Hairy Root. **Agrobacterium rhizogenes**, NE.
BACTERIAL Leaf Spot. **Pseudomonas syringae** pv. **mori**, general.
BACTERIAL Scorch. **Xylem limiting bacteria**, Mid-Atlantic and Southern U.S; **Xylella fastidiosa**, CA.
BLIGHT, Berry; Popcorn Disease. **Ciboria carunculoides**, NC to FL, TX, VA.
BLIGHT, Twig. **Myxosporium diedickei**, TX, WA.
CANKER; Twig Blight. **Cytospora** sp., NJ, TX; **Dothiorella** sp.; **D. mori**, NJ, TX; **Gibberella baccata** var. **mori**, widespread; **Nectria** sp., widespread; **N. cinnabarina**, wide spread; **Sclerotinia** sp., TX.
LEAF SPOT. **Cercospora moricola**, PA to FL, NE, TX; **Cercosporella mori**, NE, OK, TX; **Exosporium** sp., FL; **Mycosphaerella arachnoidea**, false mildew, GA, NC; **M. mori**, widespread; **Phyllosticta moricola**, KS; **Cytospora** sp., VA.
NEMATODE, Root Knot. **Meloidogyne** spp., NJ, NC, OK.
POWDERY MILDEW. **Phyllactinia corylea**, OH; **Uncinula geniculata**, NY to KS, AL.
ROT, Heart. **Polyporus farlowii**, AZ, NM; **P. hispidus**, CT; **Ganoderma applanatum**; **Hymenochaete agglutinans**, MD.
ROT, Root. **Armillaria mellea**, DE, NC, OK; **Helicobasidium purpureum**, TX; **Phymatotrichum omnivorum**, TX.
ROT, Wood. **Schizophyllum commune**, CA; **Stereum cinerescens**, MA.
RUST. **Cerotelium fici** (II), LA.

Bacterial leaf spot damages nursery trees having overhead irrigation; Mycosphaerella leaf spot sometimes defoliates older trees.

MUSHROOM, OYSTER (*Pleurotus ostreatus*)
DRY BUBBLE. **Verticillium fungicola**, PA.

MUSTARD GREENS (*Brassica juncea*)
BACTERIAL Black Rot. **Xanthomonas compestris**, FL, OH.
BACTERIAL, MLO. **Aster Yellows**.
BACTERIAL, Yellows. **Spiroplasma citri**, IL.
CLUB ROOT. **Plasmodiophora brassicae**, CA, CT, OH, TX, WA.
DAMPING-OFF. **Rhizoctonia solani**, cosmopolitan.
DOWNY MILDEW. **Peronospora parasitica**, CT, FL, IA, TX.
LEAF SPOT. **Cercospora brassicicola**, IN, LA, NJ; **Cercosporella brassicae**, CA, VA; **Alternaria brassicae**, CA, black spot; **Pseudocercosporella capsellae**, CA, white spot.
NEMATODE, Root Knot. **Meloidogyne** sp., FL, MO, TX; **Heterodera schachtii**, UT.

Powdery Mildew. **Erysiphe polygoni**, AZ, CA, FL, TX.
Rot, Crown; Drop. **Sclerotinia sclerotiorum**, TX.
Rot, Root; Damping-off. **Aphanomyces raphani**, WI.
Rust. **Puccinia aristidae** (0, I), CO; II, III on grasses.
Virus. **Cauliflower Mosaic**; **Turnip Mosaic**; **Tobacco Stread**, CA.
Wilt. **Fusarium** sp.

N

NECTARINE (*Prunus persica* var. *nectarina*)
BACTERIAL Canker. **Pseudomonas syringae**, CA.
BACTERIAL Crown Gall. **Agrobacterium tumefaciens**, MO.
BACTERIAL Leaf Spot; Canker. **Xanthomonas pruni**, OK.
BACTERIAL, MLO. **Peach X-Disease**.
CANKER. **Valsa leucostoma**, DC.
LEAF CURL. **Taphrina deformans**, CA, OR, TX, WA.
LEAF SPOT. **Cristulariella pyramidalis**, FL.
LEAF SPOT; Shot Hole. **Coryneum carpophilum**, CA, OR, WA.
NEMATODE, Lesion. **Pratylenchus thornei**, CA.
NEMATODE, Root Knot. **Meloidogyne** sp., CA; usually resistant.
POWDERY MILDEW. **Podosphaera oxyacanthae**, ID, WA; **Sphaerotheca pannosa**, ID, NY, WA.
ROT, Brown; Twig Blight. **Monilinia laxa**, CA, WA; **M. fructicola**, CT, NY, TX.
ROT, Pink Mold. **Trichothecium roseum**, CA.
ROT, Sour. **Geotrichum candidum**, CA; **Issatchenkia scutulata**, CA; **Kloeckera apiculata**, CA.
SCAB. **Cladosporium carpophilum**, CT, DE, IL, NY, PA, TX.
VIRUS. **Peach Mosaic**; **Peach Yellows**; **Peach Latent Mosaic Viroid**, WA; **Hop Stunt Viroid**, Ontario, Canada.

NEW ZEALAND SPINACH (*Tetragonia*)
BACTERIAL, MLO. **Aster Yellows**, NJ, NY.
LEAF SPOT. **Cercospora tetragoniae**, IN, MA; **Helminthosporium** sp., TX.
NEMATODE, Root Knot. **Meloidogyne** sp., TX.
VIRUS. **Beet Curly Top**, CA; **Mosaic**, DE, and **Rosette**, IN, unidentified.

R.K. Horst, *Field Manual of Diseases on Fruits and Vegetables*,
DOI 10.1007/978-94-007-5974-9_15, © Springer Science+Business Media Dordrecht 2013

O

OKRA (*Hibiscus esculentus*)

Anthracnose. **Colletotrichum gloeosporioides**, pod spot, FL, PA; **C. hibisci**, dieback, TX.

Blight, Blossoms. **Choanephora cucurbitarum**, FL, GA, TX.

Leaf Spot. **Ascochyta abelmoschi**, GA, MD, NJ, NY, pod spot; **Alternaria** sp., secondary, FL, OH, PA, SC, UT; **Cercospora abelmoschi** (*C. hibisci*), NC to FL, TX; **C. malayensis**, VA to FL, OK, TX; **Corynespora cassiicola**; **Phyllosticta hibiscina**, AL, IL, NJ, NC, OK.

Nematode, Root Knot. **Meloidogyne incognita**, general.

Powdery Mildew. **Golovinomyces cichoracearum**, CT, NJ, NC, PA.

Rot, Charcoal. **Macrophomina phaseoli**, TX.

Rot, Pod. **Botrytis** sp., NY.

Rot, Root. **Phymatotrichum omnivorum**, TX; **Rhizoctonia solani**, damping-off, AL, FL; **Thielaviopsis basicola**, NJ.

Rot, Stem. **Sclerotinia sclerotiorum**, MA.

Virus. **Tobacco Ring Spot**, GA, VA; **Yellow Mottle**, Mexico; **Yellow Mosaic Virus**, TX.

Wilt. **Fusarium oxysporum** f. sp. **vasinfectum**, CT to FL, AZ, TX; **Verticillium albo-atrum**, widespread.

OLIVE (*Olea*)

Anthracnose. **Gloeosporium olivarum**, CA.

Bacterial Knot. **Pseudomonas syringae** pv. **savastanoi**, CA.

Black Mildew; Leaf Spot. **Asternia oleina**, FL, GA.

Dodder. **Cuscuta indecora**, CA.

Leaf Spot. **Cycloclonium oleaginum**, peacock spot, CA; Cercospora caldosporioides, also fruit spot, CA.

Nematode, Citrus. **Tylenchulus semipenetrans**.

Nematode, Lesion. **Pratylenchus musicola** (*P. coffeae*), CA.

Nematode, Root Knot. **Meloidogyne** sp., CA.

Nonparisitic. **Bitter Pit** and **Dry Rot** of fruit. Overnutrition.

R.K. Horst, *Field Manual of Diseases on Fruits and Vegetables*,
DOI 10.1007/978-94-007-5974-9_16, © Springer Science+Business Media Dordrecht 2013

Exanthema. Dieback. Deficiency of organic matter.
Fruit Pit. Boron deficiency.
Soft Nose. Blue Nose. Moisture supply. On variety Sevillano only.
ROT, Root. **Armillaria mellea**, CA, TX; **Phymatotrichum omnivorum**, TX; **Polyporus olaea**, CA; **Fusarium solani**, Argentina.
ROT, Root, Twig Wilt. **Phytophthora nicotianae**, Argentina.
VIRUS. **Sickle Leaf.**
WILT. **Verticillium albo-atrum**, CA.

ONION (*Allium cepa*)
BACTERIAL Leaf Blight. **Xanthomonas campestris**, CA, FL, HI, TX; **X. axonopodis**, GA; **Pantoea ananatis**. CO.
BACTERIAL, MLO. **Aster Yellows** and **California Aster Yellows**, widespread.
BACTERIAL Rot; Slippery Skin. **Pseudomonas alliicola**, MA, NY, WA; **P. cepacia**, sour skin, scale rot, NY; **Enterobacter cloacae**, bulb decay, CO.
BACTERIAL Soft Rot. **Erwinia carotovora**, widespread in field, transit, storage.
BACTERIAL, Yellows. **Spiroplasma citri**, CA.
BLIGHT, Southern. **Sclerotium rolfsii**, AL, CA, GA, OK, TX.
BLOTCH, Purple, Brown. **Alternaria porri**, ME to MS, CO, TX, UT; **A. tennis**, CO; **A. tenuissima**, CO.
BULB DISEASE. **Panatoea ananatis**, NY.
DAMPING-OFF. **Pythium** spp., ID, MA, NY; **Rhizoctonia solani**, occasional.
DECAY, Bulb; internal. **Enterobacter cloacae**, CA.
DODDER. **Cuscuta** spp., widespread.
DOWNY MILDEW. **Peronospora destructor**, general.
LEAF BLIGHT; Bulb Rot. **Pantoea agglomerans**, GA.
LEAF NECROSIS; Seedling Death. **Phytophthora nicotianae**, NM.
LEAF SPOT. **Heterosporium allii**, CA, CO, WA; **Phyllosticta allii**, IL, NM; **Stemphylium vesicarium**, TX.
NEMATODE, Bulb. Onion Bloat. **Ditylenchus dipsaci**, NY, TX.
NEMATODE, Root Knot. **Meloidogyne incognita**.
NEMATODE, Sting. **Belonolaimus gracilis**.
NEMATODE, Stubby Root. **Paratrichodorus christiei**.
NONPARASITIC. **Blast**, of inflorescence. Appears in Connecticut Valley when bright sun follows cloudy weather. Air pollution in NJ.
Chlorosis. Copper deficiency, FL, NY; Manganese deficiency, NY, RI.
Scald. High temperature, general in summer.
Stain, alkali spot. Contact with alkaline materials or ammonia fumes in transit.
PHYTOPLASMA. **Aster Yellows, Clover Proliferation**, TX.
POWDERY MILDEW. **Oidiopsis taurica**, CA; **Leveillula taurica**, ID, Pacific Northwest.
ROT, Basal. **Fusarium oxysporum** f. sp. **cepae**, other species, widespread.
ROT, Black Stalk, Tip Blight, Seed Mold. **Stemphylium botryosum**, AR, CA, LA, NH, TX, WA.
ROT, Bulb. **Enterobacter cloacae**, WA.

ROT, Charcoal. **Macrophomina phaseoli**, CA, OK, TX. ROT, Dry. **Diplodia natalensis**, TX; **Helminthosporium allii**, also canker.

ROT, Neck; Gray Mold. **Botrytis allii**, CA, LA, NJ, TX; **B. cinerea**; Small Sclerotial; Leaf Blight, **B.** (*Botryotinia, Sclerotinia*) **squamosa**, occasional; **B. porri**, WA.

ROT, Pink Root. **Pyrenochaeta terrestris**, widespread.

ROT, Root. **Thielaviopsis basicola**, TX.

ROT, Smudge. **Colletotrichum circinans**, general.

ROT, Soft. **Kluyveromyces marxianus** var. **marxianus**, OR, WA.

ROT, Various, of bulbs. **Aspergillus niger**, black mold, general in market; **Penicillium** spp., blue mold; **Rhizopus stolonifer**, soft, after sunscald or freezing; **Sclerotinia sclerotiorum**, watery soft; **Fusarium proliferatum**, bulb rot, WA.

ROT, White. **Sclerotium cepivorum**, CA, KY, LA, ME, NJ, NY, OH, OR, PA, TX, VA.

RUST. **Puccinia asparagi** (0, I, II, III), CA, CT, IA, KS, MN; **P. allii** (*P. porri*) (II, III), CA, CT, NE.

SMUT. **Urocystis cepulae**, general; **U. colchici**.

TWISTER DISEASE. **Colletotrichum gloeosporium**, GA.

VIRUS. **Onion Yellow Dwarf**; Scape Blight caused by **Impatiens Necrotic Spot** and other Tospovirus isolate/strain, ID, including Iris Yellow Spot, CO, GA, ID, NM, OR, TX, WA (also on wild onions, **Allium pskemense**, **A. vavilovii**, and **A. altaicum**); **Iris Yellow Spot,** CA, CO, HI, GA, NY, NV, NM, OR, TX, WA, Ontario, Canada, Uruguay.

Smut is the most general onion disease, but it seldom afflicts onions grown from sets, the usual method for a small garden. Growing colored instead of white onions avoids smudge and neck rots to some extent. Sweet Spanish onions are resistant to pink root and yellow dwarf.

ORANGE
See Citrus Fruits.

P

PAPAYA (*Carica*)

ANTHRACNOSE. **Colletotrichum capsici**, FL.

ANTHRACNOSE; Fruit, Stem, and Leaf Spot. **Glomerella cingulata**, general.

BLACK MILDEW. **Asterina caricarum**, FL.

BLIGHT, Flower; Leaf Spot. **Choanephora americana**, FL.

BLIGHT, Southern. **Sclerotium rolfsii**, TX.

DAMPING-OFF. **Pythium aphanidermatum**, CA, FL; **P. debaryanum**, TX; **Colletotrichum gloeosporioides**, HI; also seedling blight.

LEAF SPOT. **Asperisporium caricae**, FL; **Mycosphaerella caricae**, target spot, FL; **Phyllosticta caricae-papayae**, target spot, FL.

POWDERY MILDEW. **Golovinomyces cichoracearum**, CA; **Oidium caricae**, FL, TX.

NEMATODE. **Meloidogyne** sp., FL, TX.

ROT, Crown, Collar. **Calonectria** sp. (*Cylindrocladium*), HI.

ROT, Fruit. **Diplodia** sp., TX; **Colletotrichum magna**, Brazil.

ROT, Root. **Phymatotrichum omnivorum**, TX.

ROT, Stem. **Fusarium** sp., CA, TX.

VIRUS. **Tobacco Ring Spot**; **Papaya Mosaic**; **Papaya Ringspot**.

PARSLEY (*Petroselinum*)

BACTERIAL, MLO. **Aster Yellows**, CO, NY; **California Aster Yellows**, CA.

BACTERIAL Soft Rot. **Erwinia aroideae**, CA; **E. carotovora**, FL.

BLIGHT, Gray Mold. **Botrytis cinerea**, AK.

BLIGHT, Leaf. **Alternaria dauci**, CT; **Septoria petroselini**, CA, CT, NJ, NY.

DAMPING-OFF; Root Rot. **Pythium** sp., NJ; **Rhizoctonia solani**, NJ, NY.

DODDER. **Cuscuta** sp., TX.

LEAF SPOT. **Cercospora petroselini**, NJ; **Stemphylium** sp., NJ.

NEMATODE, Bulb. **Ditylenchus dipsaci**, CA.

NEMATODE, Pin. **Paratylenchus projectus**, MD.

NEMATODE, Root Knot. **Meloidogyne** spp., FL, GA, KS, TX, VA.

POWDERY MILDEW. **Erysiphe heraclei**, CA, FL, WA.

R.K. Horst, *Field Manual of Diseases on Fruits and Vegetables*,
DOI 10.1007/978-94-007-5974-9_17, © Springer Science+Business Media Dordrecht 2013

Rot, Root. **Phymatotrichum omnivorum**, TX; **Phytophthora cryptogea**, CA.
Rot, Stem. **Sclerotinia sclerotiorum**, CT, GA, LA, NY, PA, TX, VA.
Virus. **Alfalfa Mosaic**, CA; **Strawberry Latent Ringspot,** CA; **Beet Curly Top**, CA; **Apium Virus Y, Potyvirus,** CA.

PARSNIP (*Pastinaca*)
Bacterial Crown Gall. **Agrobacterium tumefaciens**, VA.
Bacterial, MLO. **Aster Yellows**, IL, ME, NY, PA, SD, TX, WI, and **California Aster Yellows**, CA, WA.
Bacterial Rot. **Pseudomonas pastinacae**, NY; **Erwinia carotovora**, occasional in storage.
Blight, Leaf; Canker. **Itersonilia perplexans**, GA, MA, NY.
Downy Mildew. **Plasmopara nivea**.
Leaf Spot. **Alternaria** sp., NJ; **Cercospora pastinacae**, MD, NE, NY, TX, WV; **C. pastinacina**, CA, IN, MI, NY; **Cylindrosporium pastinacae**, UT, WI; **Ramularia pastinacae**, IN, MA, NY, ND, OH; **Septoria pastinacae**, TX; **Cercosporella pastinacae**, MA; **Phomopsis dichenii**.
Nematode, Root Knot. **Meloidogyne** spp., NJ to KS and southward.
Nonparasitic. **Heart Rot**. Boron deficiency, NY.
Powdery Mildew. **Erysiphe unbelliferarum**.
Rot, Black, Crown. **Centrospora acerina**.
Rot, Black Mold. **Rhizopus** spp., occasional in storage.
Rot, Gray Mold. **Botrytris cinerea**, occasional.
Rot, Root. **Phymatotrichum omnivorum**, AZ, TX; **Rhizoctonia solani**, black scurf, NC, TX; **Phytophthora parasitica**, dry rot, PA; **Phoma** sp.
Rot, Watery Soft. **Sclerotinia sclerotiorum**, CA, IN, LA, MA.
Scab. **Streptomyces scabies**, WA.
Scab, Acid. **Streptomyces acidiscabies**, ME.
Virus. **Mosaic**, unidentified, OR, UT.

PEA (*Pisum*)
Anthracnose, Leaf and Pod Spot. **Colletotrichum pisi**, CT, GA, IA, ME, MS, TX.
Bacterial Blight. **Pseudomonas syringae** pv. **pisi**, general, especially East and South.
Bacterial Leaf Spot. **Pseudomonas syringae**, ID.
Bacterial Pink Seed. **Erwinia rhapontici**, MT, Alberta, Canada.
Blight. **Ascochyta pisi**, SD.
Blight, Ashy Stem, Charcoal Rot. **Macrophomina phaseoli**, TX.
Blight, Foot Rot. **Mycosphaerella pinodes**, widespread, usually with **Ascochyta pinodella**.
Blight, Leaf and Stem. **Choanephora conjuncta**, GA.
Blight, Southern. **Sclerotium rolfsii**, FL.
Damping-Off; Root Rot, Pod Rot. **Pythium** spp. and **Rhizoctonia solani**, general.
Dodder. **Cuscuta** sp., MO.
Downy Mildew. **Peronospora viciae** (*P. pisi*), general, especially north central and Pacific Coast states.

LEAF SPOT. **Alternaria** sp., seedling blight, DE, NH; **A. alternata**, blight, WI; **Ascochyta pisi**, general but rare in Northwest; **Cercospora pisa-sativae**, GA; **Fusicladium pisicola**, black leaf, ID, UT; **Pleospora hyalospora**, MS; **Septoria flagellifera**, MN, ND, SD, WI; **S. pisi**, widespread; **Stemphylium polymorphum**, ME.

NEMATODE, Hop Cyst. **Heterodera humuli**, OR.

NEMATODE, Lesion. **Pratylenchus neglectus, P. thornei**, ID.

NEMATODE, Oat Cyst. **Heterodera avenae**, WA.

NEMATODE, Pea Cyst. **Heterodera gottingiana**, WA.

NEMATODE, Root Knot. **Meloidogyne** spp., AZ, CA, FL, NC, SC, TX, UT, WI.

NEMATODE, Sting. **Belonolaimus gracilis**.

NONPARASITIC. **Chlorosis**. Manganese or zinc deficiency, FL, TX, WA.
Intumescence. Pod swellings, CA, NJ, WA.
Seed Spotting. Cause unknown.

PINK SEED. **Erwinia rhapontici**, ND.

POWDERY MILDEW. **Erysiphe polygoni**, general; **E. pisi, E. trifolii, E. baeumleri**, WA.

ROT, Black Mold. **Rhizopus stolonifer**, cosmopolitan; gray mold, **Botrytis cinerea**, occasional.

ROT, Foot. **Aphanomyces euteiches**, general except in far North.

ROT, Root. **Aphanomyces euteiches**, ID, MN; **Fusarium solani** f. sp. **pisi**, widespread; **Pellicularia filamentosa**, stem canker, general; **Phoma** sp., NJ, WI; **Pyrenochaeta terrestris**, IA; **Phymatotrichum omnivorum**, TX; **Phytophthora** sp., CA, CT; **Thielaviopsis basicola**, AR, CA; **Rhizoctonia oryzae**, WA.

ROT, Stem and Root. **Rhizoctonia solani**, ND.

ROT, Stem; Wilt. **Sclerotinia sclerotiorum**, CA, DE, FL, ID, MT, NJ, PA, TX, VA, WA.

RUST. **Uromyces fabae** (0, I, II, III), CA, ID, ME, MA, MN, NE, ND, WA, WY.

SCAB; Black Spot; Leaf Mold. **Cladosporium pisicola**, CA, ME, OR, TX, UT, WA.

VIRUS. **Pea Enation Mosaic**; **Pea Mosaic**; **Pea Mottle**; **Pea Streak**; **Wisconsin Pea Streak**; **Pea Wilt**; **Clover** (red) **Vein Mosaic**; **Tomato Spotted Wilt**; **Bean Leaf Roll**, CA.

VIRUS. Seed-borne; **Pea Stunt**, WI, **Turnip Mosaic Virus**, MN.

WILT. **Fusarium oxysporum** f. sp. **pisi**, NH to SC, CA, CO, ID, IL, MN, NE, MT, OR, WA; and race 2, Near Wilt; **Fusarium oxysporum** f. sp. **medicaginis**, MS.

Resistant varieties are the answer to Fusarium wilt and some virus diseases. Using clean seed, preferably grown in the West, is the best way to avoid bacterial blight. A well-drained, fertile soil, 3–5-year rotation, cleaning up or plowing under pea refuse immediately after harvest, all help to produce healthy peas. Some organic soil amendments reduce pea wilt.

PEACH (*Prunus persica*)

ANTHRACNOSE. **Colletotrichum acutatum**, SC.

BACTERIAL Crown Gall. **Agrobacterium tumefaciens**, general.

Bacterial Hairy Root. **Agrobacterium rhizogenes**, general.
Bacterial Leaf Spot. **Xanthomonas pruni**, eastern, central, southern states.
Bacterial, MLO. **Peach X-Disease**, CT, MA, MI, NY, OH, PA.
Bacterial Shoot Blight; Canker; Gummosis. **Pseudomonas syringae**, CA, OR.
Blight, Blossom; Green Fruit Rot. **Sclerotinia sclerotiorum**, CA.
Blight, Leaf. **Fabraea maculata**, CA, NJ.
Blight; Shot-Hole Disease; Pustular Spot. **Coryneum carpophilum**, general.
Blight, Twig. **Coniothyrium** sp., TX; **Cyphella marginata**, OR; **Nectria cinnabarina**, AL.
Canker, Crown. **Phytophthora citrophthora**, CA; **P. cactorum**, stem, AR, CA; **P. syringae**, OR.
Canker; Dieback; Gummosis. **Botryosphaeria ribis** var. **chromogena**, FL, GA; **B. dothidea**, AL, FL, GA, LA, TN, TX; **Valsa** (*Cytospora*) **cincta** and **V. leucostoma**, widespread East and central states; **Coniothrium fuckelii**, WV; **Ceratocystis fimbriata**, CA.
Canker; Peach; Constriction Disease. **Fusicoccum amygdali**, East and South. Sometimes reported as **Phoma persicae**.
Damping-Off. **Rhizoctonia solani**, AR, CT.
Dieback. **Phomopsis** sp., AL. Leaf Curl. **Taphrina deformans**, general.
Leaf Scorch, Stunting, Branch Dieback. **Xylella fastidiosa**, NM.
Leaf Spot. **Phyllosticta circumscissa**, widespread; **P. persicae**, MD, NE, OH; **Mycosphaerella persicae**, frosty mildew, widespread.
Leaf Spot; Shot-Hole. **Cercospora circumscissa**, general; **C. consobrina**, IL, LA.
Nematode. **Tylenchusus palustris**, AL, AR, GA, VA.
Nematode, Dagger. **Xiphinema americanum**; **X. pacificum**, GA.
Nematode, Lesion. **Pratylenchus penetrans**; **P. vulnus**.
Nematode, Ring. **Criconemoides similis**, **C. xenoplax**, NJ.
Nematode, Root Knot. **Meloidogyne incognita**; **M. javanica**, on "nematode-resistant"varieties; **M. arenaria**.
Nematode, Sting. **Belonolaimus gracilis**.
Nonparasitic. **Catface**. Insect blemish on fruit, often from plant bugs.
Chlorosis. Iron and magnesium deficiency in alkaline soils, Southwest.
Gummosis. Winter injury, bad drainage.
Internal Bark Necrosis. Manganese toxicity, IL.
Little Leaf. Zinc deficiency, CA.
Stem Pitting. Cause unknown, WV.
Suture Spot. Cause unknown. Decline, PA.
Phytoplasma. Peach Rosette. Ontario Canada.
Powdery Mildew. **Podosphaera oxyacanthae**, general; **Sphaerotheca pannosa** var. **persicae**, general.
Rot, Brown. **Monilinia fructicola**, general, SC.
Rot, Brown; Blossom and Twig Blight. **Monilinia laxa**, Pacific states.
Rot, Bud and Twig Blight. **Fusarium lateritium**, other spp., CA, GA, KS.
Rot, Fruit. **Botryosphaeria obtusa**, AL, GA; **B. dothidea**, AL, GA, SC; **Aspergillus niger**, black mold, cosmopolitan; **Botrytis cinerea**, gray mold, cosmopolitan;

Cephalothecium roseum, pink mold, widespread; **Choanephora persicaria**, in market; **Diplodia natalensis**, gumming disease; **D. persicae**, GA; **Fusarium** spp., CA, IL, NY, TX; **Gibberella persicaria**, CA; **Gilbertella persicaria**, SC; **Glomerella cingulata**, occasional, AL, GA; **Rhizopus nigricans**, cosmopolitan; **Phomopsis** sp., NJ.

ROT, Pink Mold. **Trichothecium roseum**, CA.

ROT, Root. **Armillaria mellea**, cosmopolitan; **Clitocybe monadelpha**, AR, FL, MO, OK; **C. parasitica**, OK; **C. tabescens**, FL, GA; **Ganoderma curtisii**, NC, VA; **Phymatotrichum omnivorum**, AZ, TX; **Cylindrocladium floridanum**, GA.

ROT, Root and Crown. **Phytophthora megasperma**, NY, OH; **P. cryptogea**, NY, OH; **P. cactorum**, NY, OH.

ROT, Seedling Stem. **Sclerotium rolfsii**, South; wilt, **Cylindrocladium floridanum**; **Penicillium** sp.

ROT, Silver Leaf. **Stereum purpureum**, occasional.

ROT, Sour. **Geotrichum candidum**, CA, SC; **Issatchenkia scutulata**, CA; **Kloeckera apiculata**, CA.

ROT, Wood. **Fomes** spp.; **Lenzites saepiaria**; **Polyporus hirsutus**, **P. lacteus** and **P. versicolor**, cosmopolitan; **Stereum hirsutum**; **Shizophyllum commune**, wound rot.

RUST. **Tranzschelia discolor** (II, III), general; 0, I on anemone, hepatica.

SCAB. **Cladosporium carpophilum**, general.

SOOTY MOLD; Fruit Stain. **Fumago vagans**, cosmopolitan.

VIRUS. **Peach Asteroid Spot,** OK and TX to southern CA, OR, UT, WA; **Peach Calico**, ID, WA; **Peach Dwarf**; **Peach Golden Net**, CO; **Peach Little Peach**, eastern U. S., CT to NC to MI, MO; **Peach Mosaic**, OK, TX to CA, CO, UT; **Peach Mottle**, ID; **Peach Necrotic Leaf Spot,** MI, Puebla, Mexico; **Peach Phony Disease**, Southeast to MO, TX; **Peach Red Suture**, IN, MD, MI; **Peach Ring Spot,** CA, WA, **Peach Rosette**, SC to FL, MS, TN, MI to AR, OK; **Peach Rosette Mosaic**, MI, NY; **Peach Stubby Twig**; **Peach Wart**, AZ, CA, ID, OR, WA; **Peach Yellow Bud Mosaic**, CA, **Peach Yellow Leaf Roll** (Western X-Disease, Cherry Buckskin), CA, CO, ID, OR, UT, WA; **Peach Yellows**; **Mule's Ear Disease** (Drake Almond Bud Failure); **Prunus Ring Spot**; **Cherry Yellows**; **Green Ring Mottle**; **Prunus Necrotic Ring Spot,** CA; **Tomato Ring Spot,** CA; **Peach Latent Mosaic Viroid**, WA; **Hop Stunt Viroid**, Ontario, Canada; **Plum Pox**, PA; **Stem Pitting; Fravirus carolinianan, Baccharis halimifolia,** VA.

WILT. **Verticillium albo-atrum**, occasional.

Brown rot is the No. 1 peach enemy, and spray schedules are built around it, although they start with a dormant spray for leaf curl. Get the latest advice and spray schedule from your county agricultural agent. Peach foliage is sensitive to arsenicals; do not allow a spray prepared for shade trees to touch peaches. The commercial propagator must take many precautions to provide stock free from the numerous viruses.

PEANUT (*Arachis*)

BACTERIAL Phytoplasma. **Aster Yellows**, OK.

BACTERIAL Wilt. **Pseudomonas solanacearum**, AL, FL, NC, VA.

BLIGHT. **Sclerotinia sclerotiorum**, TX.

BLIGHT, Seedling. **Rhizoctonia** spp., NC to FL, CA, OK, TX.

BLIGHT, Southern; Stem and Nut Rot. **Sclerotium rolfsii**, general.

BLIGHT, Stem. **Diaporthe sojae**, VA, WV; **Sclerotinia minor**, VA; **S. sclerotiorum**, GA; **Botrytis cinerea**, GA; **Phomopsis longicolla**, NM.

BLOTCH, Web. **Phoma arachidicola**, VA.

CANKER, STEM. **Fusarium oxysporum**, AL.

DAMPING-OFF. Synergistic interaction of **Pythium myriotylum**, **Fusarium solani**, **Meloidogyne arenaria**, FL.

LEAF SPOT. **Alternaria** sp., FL, MO, NJ, NM, SC; **Ascochyta** sp., AR; **Cercosporidium personatum**, VA; **Phoma** sp., MO, VA; **Phyllosticta** sp., AL, AR, MS; **Pleospora** sp., AR, OK; **Stemphylium** sp., ND; **Leptosphaerulin arachidicola**, GA, TX; **Sclerotinia homoeocarpa**, FL.

LEAF SPOT, Brown, Halo. **Mycosphaerella arachidicola**, general; **M. berkeleyi** (*Cercospora personata*), general.

LEAF and STEM DISEASE. **Cercospora zebrina** (clover isolate), NC.

NEMATODE. **Panagrolaimus subelongatus**, associated with shoot elongation.

NEMATODE, Leaf. **Aphelenchoides** sp.

NEMATODE, Lesion. **Pratylenchus brachyurus**, AL, FL, GA, SC, VA.

NEMATODE, Reniform. **Rotylenchus reniformis**.

NEMATODE, Ring. **Criconemoides cylindricum**.

NEMATODE, Root Knot. **Meloidogyne arenaria**; **M. hapla**; **M. javanica**, GA.

NEMATODE, Spiral. **Helicotylenchus nannus**.

NEMATODE, Sting. **Belonolaimus gracilis**; **B. longicaudatus**. NEMATODE, Stubby Root. **Trichodorus christiei**.

NEMATODE, Stunt. **Tylenchorhynchus claytoni**.

NONPARASITIC. **Blue Stain**. Seedcoat discoloration.

Chlorosis. Excess lime; magnesium, manganese or iron deficiency.

Necrotic Spot. Nutrient deficiency.

Pops. Empty pods. Nutritional deficiency.

Pouts. Stunting and chlorotic spotting from thrips.

ROT, Black Mold, of Pods. **Cladosporium herbarum**, cosmopolitan.

ROT, Blue Mold, of Pods and Nuts. **Penicillium** sp., cosmopolitan.

ROT, Charcoal. **Macrophomina phaseoli**, CO, NC, OK, SC, TX.

ROT, Collar. **Diplodia gossypina**; **Lasiodiplodia theobromae**, NC, VA.

ROT, Crown. **Aspergillus niger**, GA, NM, TX.

ROT, Cylindrocladium Black Rot. **Cylindrocladium crotalariae**, NC, VA; **C. parasiticum**.

ROT, Gray Mold, Leaf. **Botrytis cinerea**, CT, MD, MS, TN, VA.

ROT, Peg, Pod, Root. **Calonectria** (*Cylindrocladium*) **crotalariae**, GA.

ROT, Root. **Curvularia inaequalis**, SC; **Helminthosporium** sp., OK; **Phymatotrichum omnivorum**, AZ, TX; **Pythium** sp., CA, GA, NC; **Thielaviopsis basicola**, NC.

ROT, Root and Pod. **Sclerotinia minor**, **S. sclerotiorum**, NC, VA; **Sclerotium rolfsii**, FL; **Sclerotinia sclerotiorum**, GA.

Rot, Seed. **Rhizopus** spp.; **Trichoderma viride**, SC.
Rot, Stem, Pod. **Fusarium** spp., also root rot, wilt; **Physalospora rhodina**, FL, GA; **Rhizoctonia solani**; **Sclerotium rolfsii**, FL; **Sclerotinia sclerotiorum**, GA, NM; **S. minor**, TX.
Rust. **Puccinia arachidis** (II), occasional, FL, GA, NC, TX, VA.
Virus. **Peanut Stunt**, AL, FL, GA, NC, VA; **Peanut Mottle**, GA, OK; **Cowpea Chlorotic Mottle**, SD; **Peanut Stripe**, GA; **Zuccinia Yellow Mosaic**, GA; **Impatiens Necrotic Spot**, GA, TX; **Peanut Top Paralysis**, OK; **Tomato Spotted Wilt**, AL, GA, VA, TX; **Peanut Chlorotic Ringspot**, GA.
Wilt. **Pythium myriotylum**, VA; **Verticillium** sp., NM.

PEAR (*Pyrus*)
Anthracnose, Northwestern. **Neofabraea malicorticis**, OR, WA.
Bacterial Blossom, Twig Blight; Canker. **Pseudomonas syringae**, AR, CA, CT.
Bacterial Crown Gall. **Agrobacterium tumefaciens**, general.
Bacterial Fire Blight. **Erwinia amylovora**, general.
Bacterial Fruit Rot. **Erwinia carotovora**, MA.
Blight, Leaf; Black Fruit Spot. **Fabraea maculata**, general.
Blight, Thread. **Pellicularia koleroga**, NC to FL, TX, WV.
Blight, Twig. **Corticium salmonicolor**, FL, LA; **Fusarium** spp., occasional; **Phomopsis ambigua**, widespread; **Valsa leucostoma**, WA.
Blotch, Sooty. **Gloeodes pomigena**, eastern states to OK, TX.
Canker. **Botryosphaeria dothidea**, AL; **Cytospora** spp., OR, VA, WA; **Nectria cinnabarina**, coral spot; dieback; **N. galligena**, trunk canker; **Nummularia discreta**, DE, IA.
Canker, Bark. **Glutinium microsporum**, OR; **Helminthosporium papulosum**, blister canker, black pox; **Myxosporium corticola**, NY to MI, MS, OR.
Canker, Felt Fungus. **Septobasidium** spp., NC to FL, TX.
Canker, Perennial. **Neofabraea perennans**, OR, WA.
Decline, Pear. **Candidatus Phytoplasma pyri**, Ontario, Canada.
Fruit Spot; Flyspeck. **Leptothyrium pomi**, eastern states.
Leaf Spot. **Cercospora minima**, FL to TX; **C. pyri**, MI; **Coniothyrium pyrinum**, MA to AL, IA, TX; **Coryneum foliicola**, IN; **Hendersonia cydoniae**, NY; **Phyllosticta pyrorum**, IL, MS, SC; **Mycosphaerella sentina**, ashy leaf spot, fruit spot, widespread, especially in East.
Mistletoe. **Phoradendron serotinum** (**flavescens**), AZ, TX.
Mistletoe, European. **Viscum album**, CA.
Nematode, Lesion. **Pratylenchus minyus**; **P. pratensis**, CA.
Nematode, Root Knot. **Meloidogyne** sp., CA.
Nonparasitic. **Bitter Pit**. Moisture irregularity, Pacific Coast, NY.
Black End; **Hard End**. Oriental pear rootstocks on shallow, poor soil.
Black Leaf; **Brown Bark Spot**; **Brown Blotch**, of Kieffer fruit. Undetermined.
Chlorosis. Mineral deficiency, soil alkalinity, Pacific Coast.
Cork; Drought Spot; Fruit Pitting. Boron deficiency, Pacific Coast, TX.
Exanthema. Copper deficiency, CA, FL.

Little Leaf; Rosette. Zinc deficiency, sometimes boron, CA.

Marginal Leaf Blight; Scorch. Potassium or calcium deficiency, ID, WA.

Red Leaf. In Oriental pear, undetermined cause.

Scald. Immaturity; deficient ventilation.

Stigmonose. Insect punctures during growth, widespread.

Target Canker, Measles. Undetermined, CA, GA, NY, VA, WA.

Postharvest Disease. **Sphaeropsis pyriputrescens**, MI.

Powdery Mildew. **Podosphaera leucotricha**, CO, MD, OR, WA; **P. oxyacanthae**, NJ.

Rot, Bitter; Twig, Branch Canker. **Glomerella cingulata**, widespread but not destructive.

Rot, Black; Canker; Leaf Spot. **Physalospora obtusa**, widespread.

Rot, Brown. **Monilinia fructicola**, eastern states, TX, WA.

Rot, Brown; Blossom Blight. **Monilinia laxa**. Pacific states.

Rot; Blossom and Twig Blight. **Botrytis cinerea**, widespread.

Rot, Collar, Root. **Phytophthora cactorum**, widespread; **P. citrophthora**, CA.

Rot, Fruit. **Alternaria** sp., black mold; **Aspergillus** sp., cosmopolitan; **Botryosphaeria ribis**, black rot, VA; **Cephalosporium carpogenum**, storage; **Trichothecium roseum**, occasional pink mold; **Cladosporium** sp., occasional; **Coprinus urticicola**, OR; **Gloeosporium** sp., widespread **Neurospora sitophila**, ripe rot, NC; **Penicillium** spp., blue mold, widespread; **Phialophora malorum**, storage; **Phoma exigua**; **P. mali**; **Rhizopus nigricans**, black mold, cosmopolitan; **Pleospora fructicola**; **Sclerotinia sclerotiorum**, CA, WA; **Sporotrichum malorum**, storage rot, OR, WA; **Phacidiopycnis washingtonensis** (also Canker and Blight), WA; **Sphaeropsis pyriputrescens**, WA; **Potebniamyces pyri**, WA.

Rot, Heart. **Fomes igniarius**; **F. pinicola**; **Polyporus** spp.

Rot, Root. **Armillaria mellea**, widespread; **Phymatotrichum omnivorum**, AZ, OK, TX; **Xylaria** sp., ID, IN; **Xylaria mali**, VA; **Phytophthora cinnamomi**.

Rot, Silver Leaf. **Stereum purpureum**, NY, OR; **S. hirsutum**, trunk rot.

Rot, Stem, Calyx. **Phacidiopycnis piri** (also canker and postharvst fruit rot), OR.

Rot, Trunk. **Schizophyllum commune**, cosmopolitan.

Rust. **Gymnosporangium clavariiforme** (0, I), leaves and fruit, SC: **G. clavipes** (0, I), chiefly on fruit; **G. fuseum**, CA; **G. globosum** (0, I), on leaves and fruit, eastern states to IA, MN; **G. hyalinum** (0, I), FL; **G. kernianum** (0, I), on leaves and fruit, AZ, CO; **G. nootkatense** (0, I), on Asiatic pear, AK; **G. sabinae**, MI.

Scab. **Venturia pyrina**, general.

Spot Anthracnose, Scab. **Elsinoë pini**, OR, WA.

Viroid. **Pear Blister**. Argentina.

Virus. **Apple Chlorotic Leaf Spot, Spy Epinasty Decline**, **Virginia Crab Stem Pitting and Brownline**, **Apple Mosaic**, **Flat Limb**, **Pear Vein Yellows/Red Mottle, Pear Ring-Pattern Mosaic**, **Pear Latent** on Bradford Pear (*Pyrus calleryana*), MD; **Pear Stony Pit**, Pacific Coast states, NY; **Pear Decline**, CA, OR, WA; **Pear Leaf Curl**, CA; **Pear Bark Measles**, CO; **Apple Stem Pitting**, Argentina.

Fire blight is the limiting factor in pear production, many orchards having been abandoned because of this devastating bacterial disease. Kieffer pears and some Asiatic varieties are resistant and are being used in breeding; some resistant forms are now available. Sprays during blossoming, usually an antibiotic, sometimes combined with weak copper, are used in conjunction with sanitary measures for fire-blight control. Pear decline, a virus disease transmitted by the pear psylla, is serious on the Pacific Coast.

PECAN (*Carya illinoensis*)
ANTHRACNOSE. **Glomerella cingulata**, GA.
BACTERIAL Crown Gall. **Agrobacterium tumefaciens**, widespread.
BLIGHT, Thread. **Pellicularia koleroga**, FL, NC.
BLOTCH, Leaf. **Mycosphaerella dendroides**, South.
CANKER. **Cytospora** sp., AZ.
CANKER; Black Bark Spot. **Myriangium tuberculans**, GA, MS.
CANKER; DIEBACK. **Botryosphaeria berengeriana**, AZ, SC, to FL.
CANKER, Stem. **Microcera** (*Fusarium*) **coccophila**, LA, TX.
LEAF SCORCH, FUNGAL. **Xylella fastidiosa**, **Pestalotia**, **Epicoccum**, **Curvularia**, **Fusarium**, LA.
LEAF SPOT. **Microstroma juglandis**, GA to TX; **Phyllosticta convexula**, OK; **Septoria caryae**, TX; **Pestalotia uvicola**, FL, TX.
LEAF SPOT, Algal. **Cephaleuros virescens**, FL.
LEAF SPOT, Brown. **Cercospora fusca**, prevalent through pecan belt.
LEAF SPOT; Downy Spot. **Mycosphaerella caryigena**, GA and FL to TX.
LEAF SPOT; Liver Spot. **Gnomonia caryae** var. **pecanae**, AL, AR, GA, LA, MS, TX.
LEAF SPOT; Vein Spot. **Gnomonia nerviseda**, AR, IL, LA, MS, TX.
LEAF SPOT; Zonate. **Cristulariella pyramidalis**, AL.
MISTLETOE. **Phoradendron serotinum** (**flavescens**), widespread Gulf states.
NEMATODE. **Paratylenchus** sp.; **Trichodorus** sp.
NEMATODE, Awl. **Dolichodorus obtusus**.
NEMATODE, Dagger. **Xiphinema americanum**, Southeast.
NEMATODE, Root Knot. **Meloidogyne** spp.; **M. incognita**; **M. partityla**, FL, GA.
NONPARASITIC. **Black Pit**; **Kernel Spot**. Feeding punctures of plant bugs.
Leaf Scorch. Low fertility and soil moisture capacity.
Rosette. Zinc deficiency.
Sand Burn of Seedlings. High temperature.
POWDERY MILDEW. **Microsphaera alni**, occasional.
ROT, Heart. **Schizophyllum commune**, after drought injury, OK.
ROT, Kernel and Shuck. **Phytophthora cactorum**, GA.
ROT, on nuts. **Aspergillus chevalieri**, storage mold; **Trichothecium roseum**, pink mold.
ROT, Root. **Armillaria mellea**, CA; **Helicobasidium purpureum**, TX; **Phymatotrichum omnivorum**, TX; **Pestalotia uvicola**, FL, TX; **Clitocybe tabescens**, GA.

Scab. **Cladosporium effusum**, general.

Spot Anthracnose; Nursery Blight. **Elsinoë randii**, southeastern and Gulf states.

Virus. **Bunch Disease**. Undetermined.

Pecan scab is the limiting factor in nut production in southeastern states, with strains of the fungus sometimes attacking varieties long considered immune; four or five sprays are required for control. In Arizona severe infection from Phymatotrichum root rot has come where trees are intercropped with lucerne, which provides a rapid transit medium for the fungus. Getting rid of the lucerne and treating with ammonium sulfate saves some trees. Zinc sulfate, added to the soil or to foliar sprays, controls rosette.

PEPPER (*Capsicum*)

Anthracnose; Fruit, Leaf and Stem Spot. **Gloeosporium piperatum**, MA to FL, IL, TX; **Glomerella cingulata**, fruit rot, CT to FL, KS, OH, TX; **Colletotrichum capsici**, ripe rot, MS; **C. boninense**, Brazil.

Bacterial Canker, Vascular. **Clavibacter michiganense**, IN, OH, WY, on *C. annuum* var. *cerasiforme*, and on *C. frutescens* (Bell Pepper), CA.

Bacterial Fruit and Stem Spot; Seedling Blight. **Xanthomonas vesicatoria**, general in South and East.

Bacterial Leaf Spot. **Xanthomonas vesicatoria**, NC, OH; **X. vitians**, OH; **Pseudomonas syringae**; **P. viridiflava**, FL.

Bacterial Soft Rot. **Erwinia aroideae**, and **E. carotovora** pv. **carotovora**, CA, FL, NJ; **Pseudomonas marginalis**, CA, FL, NJ, occasional in market.

Bacterial Wilt. **Pseudomonas solanacearum**, PA to FL.

Blight; Blossom Rot. **Choanephora cucurbitarum**, FL, NC.

Blight; Fruit Rot. **Phytophthora capsici**, CA, CO, FL, LA, MO, NM, NY, OH, TX, VA; **P. parasitica**, IL, IN.

Blight, Southern. **Sclerotium rolfsii**, NC to FL.

Damping-Off; Stem and Root Rot. **Pythium** spp., cosmopolitan; **Rhizoctonia solani**, cosmopolitan.

Dodder. **Cuscuta** sp., GA, NJ, VA.

Downy Mildew. **Peronospora tabacina**, GA, NC, SC, TX.

Fruit Spot, Black. Nonparasitic, TX.

Girdling, Stem. **Nectria haematococca**, FL.

Leaf Spot. **Ascochyta capsici**, WA, NY to FL; **Cercospora capsici**, frogeye; stem-end rot; **C. unamunoi**, CA and Gulf states; **C. apii**, Brazil; **Stemphylium solani**, FL; **S. botryosum** f. sp. **capsicum**, NY.

Mold, Leaf. **Cladosporium herbarum**, CA, GA, TX; Seed Mold, **Stemphylium botryosum**, CT, FL.

Necrosis, Vascular. **Verticillium dahliae**, NM.

Nematode, Awl. **Dolichodorus heterocephalus**.

Nematode, Burrowing. **Radopholus similis**.

Nematode, Lesion. **Pratylenchus penetrans**.

Nematode, Root Knot. **Meloidogyne arenaria**; **M. incognita**; **M. hapla**.

NEMATODE, Spiral. **Helicotylenchus erythrinae**.
NEMATODE, Stylet. **Tylenchorhynchus claytoni**; **T. capitatus**.
NEMATODE, Tobacco Cyst. **Heterodera tabacum**, VA.
NONPARASITIC. **Blossom-End Rot**. Deficient water supply, also lack of calcium. **Sunscald**. High temperature, often after defoliation.
PHYTOPLASMA. **Brote Grande, Phyllody**. NM, U.S. Desert Southwest.
PHYTOPLASMA, Chlorosis, Pale Green. **Candidatus Liberibacter solanacearum**, Mexico.
PHYTOPLASMA, Leaf Cup Stunt. **Candidatus Liberibacter solanacearum**, New Zealand.
PHYTOPLASMA, Little Leaf. **Phytoplasma asteris**, Mexico.
PHYTOPLASMA, Stunting, Small Leaves, Aborted Seed Pods, Costa Rica.
POWDERY MILDEW. **Leveillula taurica**, AZ, CA, FL, HI, MD, NV, NY, OK, UT, Pacific Northwest, British Columbia, Canada.
ROT, Black; Internal Mold, Early Blight. **Alternaria** spp., general after sunscald and blossom-end rot.
ROT, Charcoal. **Macrophomina phaseoli**, CA, GA, KS, NJ, TX. ROT, Crown. **Sclerotinia minor**, CA.
ROT, Collar. **Fusarium solani** f. sp. **eumartii**, CA.
ROT, Fruit. **Colletotrichum capsici** and **C. nigrum**, general; **Colletotrichum gloeosporioides** and **C. coccodes**, FL; **Diaporthe phaseolorum**, MS, MO; **Phoma destructiva**, AL, DE, FL, GA, MS, NY; **Penicillium** sp., GA; **Rhizopus stolonifer**, FL, TX, WA; **Sclerotinia sclerotiorum**, also stem rot, CA, CT, FL, MA, OH; **Fusarium** spp.; **F. solani**, **Nectria haematococca**, also stem canker; **F. subglutinans**, British Columbia, Canada.
ROT, Gray Mold. **Botrytis cinerea**, occasional in market or field.
ROT, Pod. **Curvularia lunata**, FL; **Nematospora coryli**, yeast spot after plant bug injury.
ROT, Root. **Aphanomyces** sp., NJ, of seedlings; **Phymatotrichum omnivorum**, AZ, TX; **Pythium aphanidermatum**, FL; **P. myriotylum**, FL; **P. helicoides**, FL; **P. splendens**, FL; **P. arrhenomanes**, FL; **P. catenulatum**, FL; **P. irregulare**, FL; **P. graminicola**, FL; **Fusarium solani** f. sp. **eumartii,** CA.
VIRUS. **Pepper Mottle**, CA; **Pepper Strain of Alfalfa Mosaic**; **Pepper Vein-Banding Mosaic** (Potato Y virus); **Potato Virus X**; **Cucumber Mosaic**, FL; **Tobacco Mosaic**; **Tobacco Mosaic** (Jalapeno Pepper), NY; **Tobacco Etch**; **Aster Ring Spot**; **Beet Curly Top,** Mexico; **Tomato Spotted Wilt**, FL, IN, LA, OH, OR, TN; **Pepper Mild Mottle**, FL, OR, SC; **Pepper Mild Tigre**, TX; **Pepper Texas**, TX; **Serrano Golden Mosaic**, AZ; **Tomato Leaf Curl**, CA, Mexico; **Potato Virus Y** and **Tobacco Ring Spot,** TX; **Impatiens Necrotic Spot,** WA; **Tomato Yellow Leaf Curl**, Baja, CA, Guatemala, Sur, Mexico; **Tomato Chlorosis**, Brazil; Costa Rica.
WILT. **Fusarium annuum**, AZ, CO, LA, ME, MS, NJ, NM, OK, TX; **Verticillium albo-atrum**, CA, CO, CT, LA, NY, TX; **V. dahliae**, NM.

To control bacterial spot and anthracnose rotate crops, avoiding land growing potatoes, tomatoes, or eggplant the previous year, and do not grow next to other solanaceous crops. To avoid sunscald, keep fruits shaded; spray to control leaf spots to prevent defoliation. For virus diseases, obtain healthy seed or use resistant varieties. Do not start plants in greenhouses with petunias or Jerusalem cherries; do not smoke or handle tobacco near plants; control aphid vectors; control weed hosts before the crop is planted.

PERSIMMON (*Diospyros*)

ANTHRACNOSE. **Colletotrichum acutatum**, NC.

ANTHRACNOSE; Fruit, Twig, Blight. **Gloeosporium diospyri**, East and South to KS.

BACTERIAL Crown Gall. **Agrobacterium tumefaciens**, CA.

BACTERIAL, Witches' Broom. **MLO**, TX.

BLIGHT, Thread. **Pellicularia koleroga**, FL, on Japanese persimmon.

BLIGHT, Twig. **Phoma diospyri**, FL, SC; **Physalospora obtusa**, Gulf states.

BLOTCH, Leaf. **Mycosphaerella diospyri**, Gulf states.

CANKER; DIEBACK. **Diplodia natalensis**, AL, TX; **Botryosphaeria ribis**, AL, LA, MD.

FRUIT SPOT; Flyspeck. **Leptothyrium pomi**, FL; **Macrophoma diospyri**, AL, NJ, TX; **Dissoconium mali**, IA, China.

LEAF SPOT, Algal. **Cephaleuros virescens**, FL.

LEAF SPOT, Brown. **Cercospora diospyri**, IL, ME, MS, SC, TX, VA; Black, **C. fuliginosa**, AL, FL, GA, IL, MS, TX; **Pestalotia** sp., secondary; **Phyllosticta** sp., FL, IN; **Ramularia** sp., FL; **Fusicladium levieri**, CT, FL, MS.

MISTLETOE. **Phoradendron serotinum** (**flavescens**), FL, TX; **Viscum album**, CA, on *Diospyros kakai* (Japanese Persimmon).

NEMATODE, Citrus. **Tylenchulus semipenetrans**, CA.

NEMATODE, Root Knot. **Meloidogyne** sp., FL, SC, TX.

POWDERY MILDEW. **Podosphaera oxyacanthae**, TX.

ROT, Fruit. **Alternaria** sp., occasional on Japanese persimmon; **Botrytis cinerea**, gray mold; **Penicillium expansum**, blue mold, cosmopolitan; **Physalospora** spp., AL, AZ, GA, NY, TX; **Rhizopus stolonifer**, TX.

ROT, Root. **Armillaria mellea**, CA; **Phymatotrichum omnivorum**, TX.

ROT, Wood. **Daedalea ambigua**, LA, MS; **Daldinia concentrica**, LA, VA; **Fomes** spp.; **Hericium erinaceus**, LA; **Lentinus tigrinus**, LA, MS; **Pleurotus ostreatus**, LA, MS; **Polyporus** spp.; **Schizophyllum commune**, KY, TN.

WILT. **Cephalosporium diospyri**, AL, AR, FL, GA, MS, NC, OK, SC, TN, TX; **Verticillium albo-atrum**, TX.

Persimmon wilt started in Tennessee in 1933, and at the end of 5 years only 5% of the persimmons in that native stand were alive. Oriental persimmons are resistant.

PINEAPPLE (*Ananas*)

BACTERIAL, Yeast Disease. **Klebsiella** sp., **Candida** sp., Brazil.

NEMATODE, Lesion. **Pratylenchus brachyurus**.

NEMATODE, Root Knot. **Meloidogyne** sp., FL.

NONPARASITIC. **Spike**, Long Leaf, FL.

ROT, Leaf Base; White Leaf Spot. **Ceratocystis paradoxa**, FL.

VIRUS. **Pineapple Wilt-Associated**, HI.
WILT. Toxic effect of mealybug feeding.

PISTACHIO (*Pistacia*)
BLIGHT, ROT ROOT. **Rhizoctonia solani**, CA.
BLIGHT, Shoot. **Botrytis cinerea**, CA; **Sclerotinia sclerotiorum**, CA; **Botryosphaeria rhodina**, CA; **Fusicoccum** (formerly *Botryosphaeria*).
BLIGHT, Thread. **Pellicularia koleroga**, FL, TX.
LEAF SPOT. **Phyllosticta lentisci**, TX; **Septoria pistaciarum**, AZ, NM, TX; **Alternaria** sp., NM.
LEAF SPOT, Brown. **Microascus cirrhosis**, Iran.
NEMATODE. **Meloidogyne** sp., CA; **Xiphinema index**, CA.
ROT, Root. **Phymatotrichum omnivorum**, AZ, CA, TX.
ROT, Sapwood. **Pleurotus ostreatus**, CA; **Schizophyllum commune**, CA.
WILT. **Verticillium albo-atrum**, CA.

PLUM (GARDEN), PRUNE (*Prunus domestica*)
BACTERIAL Crown Gall. **Agrobacterium tumefaciens**, widespread.
BACTERIAL Fire Blight. **Erwinia amylovora**, occasional, OR, WA.
BACTERIAL Leaf Spot; Black Spot. **Xanthomonas pruni**, eastern and southern states to WI, TX.
BACTERIAL Shoot Blight; Gummosis. **Pseudomonas syringae**, CA.
BLACK KNOT. **Dibotryon morbosum**, widespread except far West.
BLIGHT, Blossom, Twig. **Monilinia laxa**, CA, OR, WA; **Botrytis cinerea**.
BLIGHT, Thread. **Pellicularia koleroga**, LA.
BLIGHT, Twig. **Diplodia** spp., secondary.
BLOTCH, Leaf. **Phyllosticta congesta**, GA, TX.
BLOTCH, Sooty; Flyspeck. **Zygophiala cryptogama, Z. wisconsinensis, Pseudocercosporella** sp., **Stomiopeltis** spp., IA.
CANKER. **Phytophthora cactorum**, CA, IN; **Ceratocystis fimbriata**, CA; **Valsa** (*Cytospora*) **leucostoma**, dieback, widespread.
LEAF CURL; Witches Broom. **Taphrina** spp., occasional; **T. communis**, plum pockets; **T. pruni**, pockets, bladder plums; **T. insititiae**.
LEAF SCORCH. **Xylella fastidiosa**, Southern CA.
LEAF SPOT; Shot Hole. **Cercospora circumscissa**, CA, FL, MA, TX, WA; **Coccomyces prunophorae**, widespread; **Coryneum carpophilum**, CA to ID and WA; **Phyllosticta circumscissa**, IA, WA; **Septoria pruni**, TX.
MISTLETOE. **Phoradendron serotinum** (**flavescens**), TX; **Viscum album**, CA, on *Prunus salicina* ("Santa Rosa" Plum).
NEMATODE, Lesion. **Pratylenchus pratensis**; **P. vulnus**.
NEMATODE, Root Knot. **Meloidogyne** spp., FL, TX.
NONPARASITIC. **Brown Bark Spot**. Arsenical poisoning, MT.
Chlorosis. Alkaline soil, mineral deficiency, CA, FL.
Exanthema. Copper deficiency, CA, FL.
Gum Spot, Drought Spot. Irregular water supply, NY, Pacific Northwest.
Little Leaf. Zinc deficiency, CA, OR, WA.

Marginal Scorch. Fluorine injury, WA.
Myrobalan Asteroid Spot. Cause unknown, occasional.
Myrobalan Mottle. Genetic abnormality, occasional in seed stocks.
Wind Scab. Caused by developing fruit rubbing against other fruit, leaves and shoots during strong wind, CA.

POWDERY MILDEW. **Podosphaera oxyacanthae** var. **tridactyla**, occasional.
POX, Plum. NY, PA.
ROT, Brown; Blossom Blight. **Monilinia fructicola**, general; **M. laxa**, MA, NHY, RI.
ROT, Fruit. **Alternaria** sp., OR; **Botrytis cinerea**, CA, WA; **Cladosporium** sp., ID, OR; **Lambertella pruni**, CA.
ROT, Heart. **Fomes applanatus**, OR; **F. fulvus**, widespread; **Lenzites saepiaria**, Pacific Northwest; **Polyporus hirsutus**; **P. versicolor**, wide-spread.
ROT, Pink Mold. **Trichothecium roseum**, CA.
ROT, Root. **Armillaria mellea**, widespread; **Phymatotrichum omnivorum**, TX.
ROT, Silver Leaf. **Stereum purpureum**, WA.
RUST. **Tranzschelia discolor** (II, III), widespread, especially in South and Pacific Coast states; 0, I on anemone; **T. pruni-spinosae** (II, III); 0, I, on anemone, hepatica, thalictrum, buttercup.
SCAB. **Cladosporium carpophilum**, widespread.
VIRUS. **Plum Line Pattern**; **Plum White Spot**; **Prune Diamond Canker; Prune Constricting Mosaic**; **Prune Dwarf**; **Cherry Vein Clearing**; **Little Peach**; **Peach Mosaic**; **Peach Rosette**; **Peach Yellows**; **Ring Spot**; **Plum Rusty Blotch**; **Apricot Ring Pox**; **Stem Pitting; Cherry Green Ring Mottle,** North America.
WILT; Seedling Black Heart. **Verticillium albo-atrum**.

Plums are even more subject to brown rot than peaches and take the same spray schedule. This includes insect control to avoid infection through feeding injuries. Black knot is sometimes conspicuous on twigs, but diseased portions can be pruned out and a dormant lime-sulfur spray applied.

POMEGRANATE (*Punica granatum*)
ANTHRACNOSE; Fruit Spot. **Colletotrichum** sp., FL.
BLIGHT, Thread. **Pellicularia koleroga**, FL.
BLOTCH, Leaf; Fruit Spot. **Mycosphaerella lythracearum**, FL to MS, TX.
DECAY. **Pilidiella granati**, CA.
NEMATODE. **Meloidogyne** sp., FL.
ROT, Fruit. **Alternaria** sp., CA; **Aspergillus niger**, AZ, CA, TX; **Botrytis cinerea**, gray mold, cosmopolitan; **Nematospora coryli**, dry rot, CA; **Penicillium expansum**, blue mold, cosmopolitan; **Coniella granati**, NC.
ROT, Root. **Clitocybe tabescens**, FL; **Phymatotrichum omnivorum**, TX.
SPOT ANTHRACNOSE. **Sphaceloma punicae**, FL, LA, TX.

POTATO (*Solanum tuberosum*)
ANTHRACNOSE; Black Dot Disease. **Colletotrichum atramentarium** (*C. coccodes*), probably general after wilt; **Gloeosporium** sp., IN, OH.

BACTERIAL Blackleg. **Erwinia carotovora** pv. **atroseptica**, general.
BACTERIAL Canker. Vascular. **Clavibacter michiganense**, WY.
Also on *Solanum ciliatum, S. gilo, S. guineense, S. khasianum, S. nigrum, S. atropurpureum, S. avioculare, S. carolinense, S. indicum, S. nodiflorum, S. quitoense.*
BACTERIAL, MLO. **Aster Yellows**, AR, Purple-Top.
BACTERIAL Ring Rot. **Clavibacter sepedonicum**, general.
BACTERIAL Soft Rot. **Erwinia carotovora** and **E. aroideae**, cosmopolitan in transit and market.
BACTERIAL Wilt. **Pseudomonas solanacearum**, chiefly in the South; **Ralstonia solanacearum**, race 3, biovar 2, WI.
BLIGHT, Early. **Alternaria solani**, general.
BLIGHT, Foliar. **Phytophthora nicotianae**, FL, MO, NE, TX.
BLIGHT, Late. **Phytophthora infestans**, general, common in Northeast, Middle Atlantic and North central states, occasional elsewhere, AK, FL.
BLIGHT, Southern. **Sclerotium rolfsii**, NC to FL, AZ, OH.
BLOTCH, Leaf. **Cercospora concors**, GA, IN, MI, NY.
DODDER. **Cuscuta** sp., DE, NE, NJ, WA.
LEAF SPOT. **Ascochyta lycopersici**, OR.
NEMATODE, Golden. **Heterodera rostochiensis** (**Globodera rostochiensis**), DE, NY, Quebec, Canada.
NEMATODE, Lesion. **Pratylenchus brachyurus**; **P. scribneri**; **P. negelectus**; **P. penetrans**, NY; **P. penetrans** and **Verticillium dahliae**, interaction (stunting and rot), WA.
NEMATODE, Pale Cyst. **Globodera pallida**, ID.
NEMATODE, Ring. **Criconemoides mutabile**.
NEMATODE, Root Knot. **Meloidogyne** spp., general; **M. arenaria**; **M. chitwoodi**, CO, TX, WA, Pacific Northwest; **M. hapla**, Pacific Northwest.
NEMATODE, Sting. **Belonolaimus gracilis**; **B. longicaudatus**.
NEMATODE, Stubby-root. WA.
NEMATODE, Tuber. **Ditylenchus destructor**, ID.
NONPARASITIC. **Black Heart**. Oxygen deficiency.
Blackening after Cooking. Drought, heat, deficient light during tuber growth; potassium deficiency, chilling.
Blackening before Cooking. Mechanical injury.
Checking. Skin rough, scruf. Partly fertilizer injury, alkalinity.
Chlorosis; Tip blight, boron deficiency; Leaf drop, magnesium deficiency.
Dimple. Depression at bud end and **Dimple skin**, pits. Cause unknown.
Elephant Hide. Fertilizer burn, on Russet Burbank.
Fasciation (tubers and aerial parts). Probably genetic.
Giant Hill. Over-sized, late-maturing plants. Genetic factors.
Glassy End. Starch deficiency; high water content.
Growth Cracks. Fluctuating moisture.
Hollow Heart. Excessive soil moisture and fertility.
Hopperburn. Marginal necrosis from leafhoppers. General.

Internal Brown Spot. Various physiological factors.
Knobbiness, "second growth". Extreme fluctuations of soil moisture.
Lenticel Enlargement. Wet soil or oxygen deficiency.
Mahogany Browning. Low temperature, ME.
Marginal Browning, bronzing. Potassium deficiency.
Ozone. Purple-black specks from air pollution.
Pitting, Spot necrosis of tubers. Oxygen deficiency in storage.
Pointed Ends. Irregular growth conditions.
Psyllid Yellows. Insect injury, western states.
Ring Spot. Zonate, depressed lesions. Cause unknown.
Scald of tubers. Overheating, sunburn, frequent in South.
Spraying. Internal concentric necrosis. TRV or PMTV Virus.
Sprout Tubers; "Little Potato". Overheating in storage; planting in dry, cold soil.
Stem-End Browning. Vascular necrosis. Virus.
Stem Necrosis; Defoliation. Manganese toxicity in acid soil.
Tipburn. Abrupt transition from cool, moist to hot, dry weather, general.

PHYTOPLASMA, Aster Yellows. **Candidatus Phytoplasma asteris**, NE, TX.

PHYTOPLASMA. **Potato Purple Top**, MT, Canada, Mexico, Russia, U.S.

PHYTOPLASMA, Zebra Chip Disease. **Candidatus Liberibacter psyllaurous**, CA, Honduras, Mexico; **C. L. solanacearum**, AZ, CA, CO, KS, NE, NV, NM, TX. Central America, Guatemala, Mexico.

POWDERY MILDEW. **Golovinomyces cichoracearum**, CA, KY, NJ; **Oidium** sp.; MD, NJ, OH, PA, UT, western U.S; **Leveillula taurica**, WA.

ROT, Charcoal. **Macrophomina phaseoli**, CA, GA, IL, OK, TX.

ROT, Collar. **Fusarium solani** f. sp. **eumartii**, CA.

ROT, Dry. **Fusarium graminearum**, ND; **F. torulosum**, **F. oxysporum** and **F. sambucinum**, MI.

ROT, Gray Mold; Shoot Blight. **Botrytis cinerea**, AK, CT, ID, ME, OH.

ROT, Pink, Watery. **Phytophthora erythroseptica**, DE, ID, LA, ME, MA, OK, PA; **P. parasitica**, TX; **P. cryptogea**, TX.

ROT, Root. **Armillaria mellea**, CA, FL, MI, OR, WA, WI, WY; **Phymatotrichum omnivorum**, TX; **Helicobasidium purpureum**, MA, MT, NE, NY, ND, OK, OR, TX, WA; **Fusarium solani** f. sp. **eumartii**, CA.

ROT, Silver Scurf. **Spondylocladium atrovirens**, general but less frequent in the South; **Helminthosporium solani**, NY.

ROT, Stem. **Sclerotinia** sp., FL, ME, SC; **S. minor**, CA; **S. sclerotiorum**, FL, MA, MT, NY, OR, TX, WA, general.

ROT, Tuber. **Aspergillus niger**; **Fusarium** spp.; **F. solani**, jelly end-rot; **Gliocladium** sp., secondary; **Oospora pustulans**, skin spot; **Penicillium** spp., blue mold; **Phoma tuberosa**; **Phomopsis tuberivora**; **Phytophthora drechsleri, P. nicotianae,** FL, MO, NE, TX; **Rhizopus** spp.; **Trichothecium roseum**, pink mold; **Xylaria apiculata**.

ROT, Tuber Sprout. **Fusarium sambucinum**, MI.

ROT, Watery Leak. Chiefly **Pythium debaryanum**, common in West; **Pythium** spp.

SCAB. **Streptomyces scabies**, general; Powdery Scab, **Spongospora subterranea**, AK, ND; Russet Scab, **Streptomyces** sp.

VIRUS. **Apical Leaf Roll**, unidentified; **Potato Aucuba Mosaic, NY**; **Potato Calico** strain of **Alfalfa Mosaic**; **Potato Crinkle**; **Mild Mosaic** (Virus X plus A); **Green Dwarf**; **Beet Curly Top**; **Potato Leaf Roll**, general; **Potato Leaf Rolling Mosaic**; **Mottle** (Potato Virus X); **Potato Rugose Mosaic** (Potato Virus Y), general; **Potato Virus Y**, Mexico; **Potato Spindle Tuber** (viroid), general; **Potato Vein Banding**; **Potato Witches' Broom**; **Potato Yellow Dwarf**; **Potato Yellow Spot**; **Tobacco Ring Spot**; **Potato Corky Ring Spot** (**Tobacco Rattle**), MI, MN, ND, WI; **Potato Virus M**; **Potato Virus S**, general; **Tobacco Veinnal Necrosis**, Pacific Northwest; **Potato Virus A**, WA; **Tomato Spotted Wilt**, NC, TX; **Potato Mop-Top**, ME, ND, WA; **Impatiens Necrotic Spot,** NY, WA**;** **Potato Yellowing**, Ecuador.

WART, Potato. **Synchytrium endobioticum**, MD, PA, VA; see under Galls; Prince Edward Island, Canada.

WILT. **Verticillium albo-atrum**, widespread.

WILT; Stem-end Rot. **Fusarium** spp.; **F. oxysporum**, general but more frequent East and South.

Potato growing is a highly specialized business not well adapted to most small gardens. Use certified seed, resistant varieties where possible. Control insects spreading virus diseases.

PUMPKIN
See Squash.

Q

QUINCE (*Cydonia*)

ANTHRACNOSE, Northwestern. **Neofabraea malicorticis**, OR, WA.

BACTERIAL Crown Gall. **Agrobacterium tumefaciens**, general.

BACTERIAL Fire Blight. **Erwinia amylovora**, general.

BACTERIAL Hairy Root. **Agrobacterium rhizogenes**, ME to NC; Pacific Coast.

BLIGHT, Dothiorella Twig. **Botryosphaeria ribis**, TX.

BLIGHT, Leaf; Black Spot. **Fabraea maculata**, general.

BLIGHT, Thread. **Pellicularia koleroga**, NC, Gulf states.

CANKER, Perennial. **Neofabraea perennans**, OR.

CANKER, Trunk. **Nectria galligena**, OR; Twig, **Valsa leucostoma**.

FRUIT SPOT. **Leptothyrella pomi**, MO; **Mycosphaerella pomi**, New England to OH.

LEAF SPOT. **Phyllosticta** sp., DE.

NEMATODE, Root Knot. **Meloidogyne** sp., TX.

POWDERY MILDEW. **Phyllactinia corylea**, VA; **Podosphaera leucotricha**, CA, WA; **P. oxyacanthae**, NY, WV to IN.

ROT, Bitter; Canker. **Glomerella cingulata**, eastern and central states southward.

ROT, Black; Canker; Leaf Spot. **Physalospora obtusa**, eastern states to AL, TX.

ROT, Brown. **Monilinia fructicola**, eastern states, MS, OR, TX; **M. laxa**, also blossom and twig blight, Pacific Coast.

ROT, Fruit. **Alternaria mali**, IN; **Botrytis cinerea**, occasional; **Cephalothecium roseum**, pink mold, occasional; **Penicillium expansum**, cosmopolitan; **Phoma cydoniae**, pale rot, IL, MI; **P. mali**, IN.

RUST. **Gymnosporangium clavariiforme** (0, I), leaves, fruit, stems, CT, ME, NH; III on juniper; **G. clavipes** (0, I), orange rust, quince rust, on fruit, stems; III on juniper; **G. gracile** (0, I); III on juniper; **G. libocedri** (0, I), fruit, stems, OR; III on incense-cedar; **G. nelsonii** (0, I), leaves, stems, AZ, CO; III on juniper; **G. nidus-avis** (0, I), leaves, fruit, stems, CT, NY; III on juniper.

SCAB. **Venturia pirina**, CT.

SPOT ANTHRACNOSE. **Elsinoë piri**, WA.

R.K. Horst, *Field Manual of Diseases on Fruits and Vegetables*,
DOI 10.1007/978-94-007-5974-9_18, © Springer Science+Business Media Dordrecht 2013

Quinces are subject to fire blight; infected branches should be cut out with the usual precautions. For leaf blight, spray when blossoms show pink, again when last of the petals are falling, and perhaps twice more at 2-week intervals. Brown rot is not very important on quince. To prevent rust, remove nearby susceptible junipers or spray them in spring as spore horns are developing on galls. See under RUSTS.

R

RADISH (*Raphanus*)

Bacterial Black Rot. **Xanthomonas campestris**, IN, IA, MI, NJ, NY, OH, PA, TX.

Bacterial, Phytoplasma. **California Aster Yellows**, CA; **Virescence**, ID, WA.

Bacterial Soft Rot. **Erwinia carotovora**, cosmopolitan.

Bacterial Spot. **Xanthomonas vesicatoria** pv. **raphani**, IN; **Xanthomonas campestris** pv. **armoraciae**, OH.

Blotch; Black Pod. **Alternaria raphani**, CA, MI, MN, NJ, OH, PA.

Club Root. **Plasmodiophora brassicae**, occasional in North, MA to NJ, MN, WA.

Damping-Off. **Pythium debaryanum**, MA, MN, NJ, WY; **Rhizoctonia solani**, cosmopolitan.

Downy Mildew. **Peronospora parasitica**, northeastern and central states to MS, CA, TX.

Leaf Spot. **Alternaria brassicae**, gray leaf spot, CA; **A. oleracea**, black leaf spot, CT, NJ; **Cercospora cruciferarum**, AL, IL, MO, TX; **C. atrogrisea**, NJ.

Nematode, Leaf and Stem. **Ditylenchus dipsaci**, NY.

Nematode, Root Knot. **Meloidogyne arenaria**; **M. javanica**, AL, MS, OR, PA, TX.

Nonparasitic. **Air Pollution**. NO2, O3, SO2, NC.

Postharvest, Black Patch. **Alternaria raphani**, St. Catharines, Canada.

Powdery Mildew. **Erysiphe polygoni**, CA, MO, TX.

Rot, Black Root. **Aphanomyces raphani**, ME to FL, CA, IA, OK, OR; **Pythium aphanidermatum**, IN, KS, MA, MI, NY, OH, OK, PA, SC, WI.

Rot, Crown, Watery Soft. **Sclerotinia sclerotiorum**, CA, IN, MN, NJ, TX.

Rot, Pod. **Phoma lingam**, CA, FL.

Rot, Root. **Ascochyta** sp.; **Phymatotrichum omnivorum**, TX. RUST. **Puccinia aristidae** (0, I), AZ, CO.

Scab. **Streptomyces scabies**, IN, MI, NJ, OH, TX, WI.

Scab, Acid. **Streptomyces acidiscabies**, ME.

Virus. **Radish Mosaic**; **Beet Curly Top**; **Tobacco Streak**, CA; **Tomato Spotted Wilt** (on wild radish), GA.

White Rust. **Albugo candida**, general.

R.K. Horst, *Field Manual of Diseases on Fruits and Vegetables*,
DOI 10.1007/978-94-007-5974-9_19,

Wilt; Yellows. **Fusarium oxysporum** f. sp. **raphani**; **F. oxysporum** f. sp. **conglutinans**, CA.

Radishes are so easily grown in home gardens that not many gardeners worry about disease control. Seed should, however, be treated for damping-off and root rots.

RAMBUTAN (*Nephelium lappaceum*)

Canker, Stem. **Dolabra nepheliae**, HI, Puerto Rico.

Rot, Fruit. **Gliocephabotrichum simplex**, HI; **G. bulbilium**, HI; **Lasmenia** sp., HI, Puerto Rico; **Pestalotiopsis virgatula**, HI.

RASPBERRY (*Rubus*)

Anthracnose; Dieback; Gray Bark. **Gloeosporium allantosporum**, OR, WA.

Bacterial Blossom Blight. **Pseudomonas** sp., OR.

Bacterial Crown Gall. **Agrobacterium tumefaciens**, general; **A. rubi**, cane gall, NY, PA to IL, OR, WI.

Bacterial Fire Blight; Flower and Twig Blight. **Erwinia amylovora**, ME, NH, PA, WA, WI.

Bacterial Hairy Root. **Agrobacterium rhizogenes**, OR.

Bacterial, Witches' Broom. **MLO**, OR.

Blight, Cane. **Leptosphaeria coniothyrium**, general; **Physalospora obtusa**, IA, MD, MI, MO, ND; **Sclerotinia** sp., NY.

Blight, Spur. **Didymella applanata**, general.

Canker; Cane Spot. **Ascospora ruborum**, MA, OR, WI.

Canker; Dieback. **Glomerella cingulata**, AR, KY, MD, MI, MO, NJ, OH, RI, WV; **G. rubicola**, white bud, IL, NJ; **Botryosphaeria ribis**, FL; **Macrophoma rubi**, IL.

Dodder. **Cuscuta gronovii**, CT, IL, WI.

Downy Mildew. **Peronospora rubi**, WA.

Fruit Spot, Flyspeck. **Leptothyrium pomi**, IN, KY, MA.

Leaf Spot. **Cylindrosporium rubi** (*Sphaerulina rubi*), also cane spot, common East of Rocky Mts.; **Mycosphaerella confusa** (*Cercospora rubi*), NJ to FL, IL, TX; **M. rubi**, general; **Pezizella oenotherae**, MD, MO, VA; **Septoria darrowi** (perhaps same as *Cylindrosporium rubi*) **Stigmatea rubicola**, black leaf and cane spot, MT, NM, NY, VT, WI.

Nematode, Lesion. **Pratylenchus vulnus**.

Nematode, Sheath. **Hemicycliophora** sp.

Nonparasitic. **Chlorosis**. Iron deficiency in West.

Powdery Mildew. **Phyllactinia corylea**, MI; **Sphaerotheca macularis**, Northeast, Northwest, CA, IL, MN.

Rosette, Double Blossom. **Cercosporella rubi**, IL, KY, MD, NY, PA.

Rot, Fruit. **Botrytis cinerea**, gray mold, cosmopolitan; **Phyllostictina carpogena**, MD; **Monilinia fructicola**, brown rot, IL; **Alternaria** sp., MA; **Rhizopus nigricans**, black mold, cosmopolitan.

Rot, Root. **Armillaria mellea**, OR, WA; **Rhizoctonia solani**, CO, ID, WA; **Xylaria** sp., WA; **Phytophthora erythroseptica**, Pacific Northwest; **P rubi**, WA; **Pratylenchus penetrans**, WA.

RUST, Late. **Pucciniastrum americanum** (II, III), Northeast to ID, IL, NC, OH; 0, I on spruce.

RUST, Leaf. **Phragmidium rubiidaei** (0, I, II, III), northeastern and central states to CO; Pacific Northwest.

RUST, Orange. **Gymnoconia peckiana** (0, I, III), on black raspberry, Northeast to MN, Pacific Northwest; **Kunkelia nitens** (I), IL, IN, MI, OH.

RUST, Yellow. **Kuehneola uredinis** (0, I, II, III), IL, PA, DE to WI.

SPOT ANTHRACNOSE. **Elsinoë veneta**, general but less common on red than black raspberry.

VIRUS. **Raspberry Leaf Curl**; **Red**; **Black Raspberry Necrosis,** TN; **Raspberry Streak**; **Raspberry Yellow Mosaic**, general; **Tomato Ring Spot,** cause of crumbly fruit, NY; **Tobacco Ring Spot,** NC; **Raspberry Bushy Dwarf**, OH; **Black Raspberry Witches' Broom Phytoplasma**, OR; **Crumbly Fruit Disease**, OR; **Rubus Yellow Net**, Alberta, Canada**; Blackberry Chlorotic Ringspot**, OR, WA.

WILT. **Verticillium albo-atrum**, MA to NJ, OH, OR, WA.

Virus diseases are important on raspberries and cannot be controlled by spraying. Purchase healthy plants and set, if possible, 500 ft away from old patches. Inspect at least three times the first year, roguing all diseased plants, after first searing them with a blow torch or flame thrower so aphids will not carry the virus to nearby healthy bushes. Plants seldom recover from Verticillium wilt and never from orange rust, which is systemic. Crown gall is important on red raspberries; if infected plants are found, raspberries should not be replanted in the same soil for several years.

RHUBARB (*Rheum*)

ANTHRACNOSE. **Colletotrichum** sp., IL, MO, PA, WV, WI.

BACTERIAL Crown Gall. **Agrobacterium tumefaciens**, IA, MA, NY.

BACTERIAL Soft Rot. **Erwinia carotovora**, occasional in market; **E. rhapontici**, crown rot, OK.

BLIGHT, Southern. **Sclerotium rolfsii**, FL, MS, TX, VA.

DAMPING-OFF; Crown Rot. **Pythium** spp., CA, MD.

DOWNY MILDEW. **Peronospora rumicis**, CA.

LEAF SPOT. **Ascochyta rhei**, eastern and central states to MS, KS; **Alternaria** sp., CA, MN, NE, NJ, PA; **Cercospora** sp., DE, MD, NE; **C. rhapontici**, IL; **Cladosporium** sp., CA, WA; **Macrophoma straminella**, general; **Ramularia rhei**, CA; **Septoria rhapontici**, IA.

NEMATODE, Root Knot. **Meloidogyne** spp., CA, MD, NY, OK.

NONPARASITIC. Crack Stem. **Boron Deficiency**, WA.

ROT, Gray Mold. **Botrytis cinerea**, serious in greenhouses.

ROT, Root. **Armillaria mellea**, CA, TX; **Phymatotrichum omnivorum**, AZ, TX; **Phytophthora cactorum**, crown rot, CA, MO, OK, PA; **P. parasitica**, IL, KS, LA, MD, MO, NY, TX, VA; **Rhizoctonia solani**, CA, CT, IL, MN, MO, NY, OK, TX, WA.

RUST. **Puccinia phragmitis** (0, I), CA, MN, NE; II, III on *Phragmites*.

Virus. **Rhubarb Chlorotic Ring**; **Turnip Mosaic**, AK; **Beet Curly Top**.

Macrophoma leaf spot is common but seldom calls for control measures beyond removal of old stalks in late fall. Plants with crown rot should be dug and burned.

ROLLINIA
Dieback, Fruit Rot. **Glomerella cingulata**, FL.

RUTABAGA
See Turnip.

S

SALSIFY (*Tragopogon*)

BACTERIAL, MLO. **California Aster Yellows**, CA; **Aster Yellows**, MD, NY, PA, WI.
BACTERIAL Soft Rot. **Erwinia carotovora**, CT, TX.
BLIGHT, Leaf. **Sporodesmium scorzonerae**, AL, MD, NY, PA, VA, WV.
BLIGHT, Southern. **Sclerotium rolfsii**, TX.
LEAF SPOT. **Cercospora tragopogonis**, MT, OK; **Stemphylium botryosum**, NY; **Alternaria tenuis**.
NEMATODE, Leaf and Stem. **Ditylenchus dipsaci**, CA.
NEMATODE, Root Knot. **Meloidogyne** sp., NY to AL, TX, WA.
POWDERY MILDEW. **Golovinomyces cichoracearum**, general.
ROT, Root. **Phymatotrichum omnivorum**, AZ, TX; **Rhizoctonia solani**, TX, WA.
ROT, Stem, Crown. **Sclerotinia intermedia**, IL; **S. sclerotiorum**, IL.
VIRUS. **Beet Curly Top**; **Lettuce Mosaic**, NY; **Tomato Spotted Wilt**, FL.
WHITE RUST. **Albugo tragopogonis**, CA.
WILT. **Verticillium albo-atrum**, NY.

SHALLOT (*Allium ascalonicum; A. porrum*, Leek)

BACTERIAL Blight. **Pseudomonas syringae** pv. **porri**, CA.
BACTERIAL, MLO. **Aster Yellows**, LA.
BLIGHT, Southern. **Sclerotium rolfsii**, LA.
BLOTCH, Purple. **Alternaria porri**, CA, LA, TX.
DOWNY MILDEW. **Peronospora destructor**, LA.
NEMATODE, Root Knot. **Meloidogyne** sp., TX.
ROT, Bulb and Basal. **Fusarium culmorum**, CA.
ROT, Neck; Gray Mold, **Botrytis allii**, LA; Smudge, **Colletotrichum circinans**, IL, LA, WI.
ROT, Pink Root. **Pyrenochaeta terrestris**, CO, LA, TX; **Fusarium solani**, TX.
ROT, White. **Sclerotium cepivorum**, LA, VA.
SMUT. **Urocystis cepulae**, MA.
VIRUS. **Mosaic**, unidentified; **Tomato Spotted Wilt**, GA; **Irish Yellow Spot**, CO, OR.

R.K. Horst, *Field Manual of Diseases on Fruits and Vegetables*,
DOI 10.1007/978-94-007-5974-9_20,

Pink root and white rot are prevalent in Louisiana. Losses from white rot are heavy if plants are set late; September setting may give a good crop.

SOYBEAN (*Glycine max*)

ANTHRACNOSE. **Glomerella glycines**, IA, MI, NC to FL, NE, TX; **Colletotrichum truncatum**; **C. coccodes**, IL.

BACTERIAL Blight. **Pseudomonas syringae** pv. **glycinea**, East and South to MN, TX.

BACTERIAL, MLO. Bud Proliferation, LA.

BACTERIAL Leaf Crinkle. Unidentified, Midwest.

BACTERIAL Pustule; Pustular Spot. **Xanthomonas glycines** (**phaseoli** var. **sojense**), general.

BACTERIAL Tan Spot. **Clavibacter flaccumfaciens**, IA.

BACTERIAL Wildfire. **Pseudomonas tabaci**, MD to AL, LA, NE.

BACTERIAL Wilt. **Pseudomonas solanacearum**, NC.

BLIGHT, Aerial. **Rhizoctonia solani**, LA.

BLIGHT, Ashy Stem; Charcoal Rot. **Macrophomina phaseoli**, NJ to SC, NE, OH, TX, WA, WI.

BLIGHT, Gray Mold; Leaf Spot. **Botrytis cinerea**, CT, OH.

BLIGHT, Leaf. **Choanephora infundibulifera**, LA.

BLIGHT, Pod and Stem. **Diaporthe sojae**, NY, MI, to GA, KS, LA.

BLIGHT. **Sclerotinia minor**, **S. sclerotiorum**, VA.

BLIGHT, Southern. **Sclerotium rolfsii**, VA to FL, IA, KS, TX.

CANKER, Stem. **Diaporthe phaseolorum** vars. **batatatis, meridionalis**, and **caulivora**, AR, FL, IL, IN, KY, LA, MD, OH, SC, TX, WI; **Phyllosticta sojicola**, AR, MD, stem and pod canker.

DAMPING-OFF. **Fusarium solani**, MN, ND.

DAMPING-OFF; Root Rot. **Pythium** spp., IL, IA, MN, MO, NC, ND; **P. aphanidermatum**; **P. torulosum**, IA; **P. delawarii**, OH; **Phytophthora megasperma** f. sp. **glycinea**, VA.

DECAY, Root, Stem. **Mycoleptodiscus terrestris**, WI.

DOWNY MILDEW. **Peronospora manshurica**, East and South to LA, IA.

LEAF SPOT. **Alternaria** sp., widespread, secondary; **Cercospora canescens**, also on pods, stems, AL, IL, MD, MS, NC, TX, WV; **C. kikuchii**, also purple stain of seed, IN, MD, NC, VA; **C. sojina**, frogeye leaf spot, MI to FL, IA, NY, OK, VA, WI, Argentina; **Corynespora cassiicola** (*Helminthosporium vignicola*), target spot, AL, AR, FL, GA, LA, MN, MS, NC, SC; **Helminthosporium vignae**, zonate leaf spot, NC; **Mycosphaerella cruenta**, GA, MS; **Myrothecium roridum**, secondary, LA; **Phyllosticta glycinea**, IL, MD, MO, NC, VA; **Pleosphaerulina sojicola**, ME, NJ, NY, WI.

LEAF SPOT, Brown Spot. **Septoria glycines**, AR, DE, IN, IA, MD, NC, WI.

LEAF and STEM DISEASE. **Cercospora zebrina** (clover isolate), NC.

MOLD, White. **Sclerotinia sclerotiorum**, ME.

NEMATODE, Cyst. **Heterodera glycines**, soybean cyst, AR, IA, KS, KY, LA, MI, MN, MO, NE, NC, ND, OH, SD, TN, TX; **H. gottingiana**, peacyst; **H. trifolii**, clover cyst; **Cactodera estonica**, WI.

NEMATODE, Dagger. **Longidorus** sp.
NEMATODE, Lesion. **Pratylenchus safaenis**, GA.
NEMATODE, Root Knot. **Meloidogyne hapla**; **M. javanica**.
NEMATODE, Spiral. **Helicotylenchus erythrinae**; **H. nannus**.
NEMATODE, Stem. **Ditylenchus dipsaci**, NY.
NEMATODE, Sting. **Belonolaimus gracilis**; **B. longicaudatus**, DE.
NONPARASITIC. **Baldhead**. Mechanical injury.
Chlorosis. Interveinal. Manganese deficiency, AL, NC, TN.
Yellowing. Potassium deficiency, iron deficiency.
PHYTOPLASMA, Stunting, Small Leaves, Aborted Seed Pods, Costa Rica.
POWDERY MILDEW. **Erysiphe polygoni**, DE, IA, NC, SC; **Microsphaera diffusa**, DE, MN, NC.
ROOT DEATH, Yellowing and Stunting. **Phytophthora sansomeana**, OR.
ROT, Charcoal. **Macrophomina phaseolina**, IA, MN, ND.
ROT, Red Crown. **Cylindrocladium crotalariae**, Telemorph, **Calonectria crotalariae**, MS.
ROT, Root. **Corynespora cassiicola**, WI; **Phymatotrichum omnivorum**, TX; **Pythium aphanidermatum**, AZ; **P. dissotocum**, AZ; **Stachybotrys chartarum**, IL; **Fusarium oxysporum, F. solani, F. acuminatum, F. proliferatum**, IA, MN; **F. redolens**, MN; **F. tricinctum**, MN.
ROT, Root Black. **Thielaviopsis basicola**, AR.
ROT, Root and Stem**. Fusarium virguliforme**, MN; **Phytophthora megasperma** f. sp. **glycinea**, DE, Northeast, MN, SD; **P. sojae**, AR, IL, MN, ND.
ROT, Seed. **Fusarium scirpi**, NE; **F. graminearum**, IL; **Nematospora coryli**, NC, OK, SC, VA; **Aspergillus** spp; **Phomopsis sojae**, NY.
ROT, Stem. **Cephalosporium gregatum**, brown stem, IL, IN, IA, KY, MN, MO, NC, OH, VA; **Pellicularia filamentosa**, also root rot, canker, general; **Sclerotinia sclerotiorum**, AZ, IA, MD, NY, VA; **Phytophthora megasperma** var. **sojae**; **Neocosmospora vasinfecta**, AL.
RUST. **Phakopsora pachyrhizi**, AL, AR, FL, GA, HI, IL, IN, IA, KY, LA, MD, MS, MO, NE, NC, SC, TN, TX, VA, Cuba, Yucatan Peninsula, Mexico; **P. packopsora**, VA, Mexico.
SEED INFECTION. **Phomopsis longicolla**, IA.
SUDDEN DEATH SYNDROME. **Fusarium solani** f. sp. **glycines**, AR, DE, IA, KY, MN, MO, NE, OH, PA; **F. tucumaniae**, Argentina; **F. virguliforme**, MI, WI; **Phoma** sp.; **Epicoccum** sp.; **Alternaria** sp., **Trichoderma** sp., MO.
VIRUS. **Soybean Mosaic**, ND; **Tobacco Ring Spot,** causing bud blight, midwestern U.S.; **Beet Curly Top**; **Bean Pod Mottle**, NE; **Soybean Yellow Mosaic**, Midwest; **Cowpea Mosaic**, IL; **Tobacco Streak**, OK; **Soybean Dwarf**, CA, IL, WI; **Tobacco Mosaic**; **Pepper Golden Mosaic**, Mexico; **Bean Pod Mottle**, AL, Ontario, Canada; **Tomato Spotted Wilt**, AL, GA; **Rhynchosia Golden Mosaic**, Mexico; **Iris Yellow, Bunyaviridae**, FL, GA; **Soybean Yellow Mottle Mosaic**, MS; **Soybean Vein Necrosis**, TN; **Alfalfa Mosaic, Alfamovirus**, TN.
WILT. **Fusarium oxysporum** f. sp. **tracheiphilum**, PA to FL, NE, TX; also CA.

Edible soybeans are well-suited to home garden culture and usually produce a good crop without control measures beyond dusting seed with a protectant before planting. Commercial growers find a number of diseases of economic importance; bacterial pustule, the various blights, leaf spots, downy mildew, wildfire, virus diseases. Some areas are now under quarantine for the soybean cyst nematode.

SPINACH (*Spinacea*)

ANTHRACNOSE. **Colletotrichum spinaciae**, CT, LA, MS, NJ, NY, TX, VA; **C. spinacicola** (*Gloeosporium spinaciae*); **C. dematium**, CA.

BACTERIAL Leaf Spot. **Pseudomonas syrinagae** pv. **syringae**, CA.

BACTERIAL, MLO. **Aster Yellows.**

BACTERIAL Soft Rot. **Erwinia carotovora**, general in transit, market.

BLIGHT, Phytophthora. **Phytophthora capsici**, IL.

DAMPING-OFF; Root Rot. **Rhizoctonia solani**, general; **Pythium** spp., preemergence seed decay.

DOWNY MILDEW. **Peronospora effusa**, general; **P. farinosa** f. sp. **spinaciae**, AZ, AR, CA, FL.

LEAF SPOT. **Cercospora beticola**, CA, GA, IL, IA, MA, NY, TX; **Phyllosticta chenopodii**, DE, NJ, NY, VA; **Stagonospora spinaciae**, SD; **Alternaria spinaciae**, MA; **Stemphylium botryosum**, AZ, CA, DE, MD; **Cladosporium variabile**, WA; **Bipolaris** sp., FL.

MOLD, Leaf. **Cladosporium macrocarpum**, secondary, DE, OK, PA, TX; **Heterosporium variabile**, pinhead "rust," general.

MOLD, Seed. **Pleospora herbarum** (*Stemphylium botryosum*); **Alternaria** sp., and secondary leaf spot; **Curvularia inaequalis.**

NECROTIC, Circular Tan Leaf. **Stemphylium botryosum**, TX.

NEMATODE. Clover Cyst. **Heterodera trifolii.**

NEMATODE, Root Knot. **Meloidogyne** spp.

NONPARASITIC. **Yellows.** Nutrient deficiency.

ROT, Root. **Aphanomyces cladogamus**, NJ, VA; **Phymatotrichum omnivorum**, TX; **Pyrenochaeta terrestris**, IA; **Olpidium brassicae**.

ROT, Root, Crown; Wilt. **Fusarium solani; F. oxysporum** f. sp. **spinaciae**, general; **Phytophthora** sp., AZ, IN, NJ, NY**; P. cryptogea**, CA; **P. megasperma**, CA, NC; **Sclerotinia sclerotiorum**, GA, NY.

RUST. **Puccinia aristidae** (0, I), AZ, CA, CO, OR, WA; II, III on grasses.

SMUT, Leaf. **Entyloma ellisii**, NJ, WA.

VIRUS. **Cucumber Mosaic** (Spinach Blight); **Spinach Yellow Dwarf**; **Beet Curly Top**; **Beet Ring Mottle**; **Beet Pseudo Yellows**; **Tomato Spotted Wilt**; **Lettuce Mosaic**, NY; **Lettuce Speckles Mottle**, CA; **Spinach Latent**; **Potato Leaf Roll**, Pacific Northwest; **Spinach Severe Curly Top,** AZ; **Impatiens Necrotic Spot**, CA; **Bean Yellow Vein**, CA; **Curtoviruses**, AZ; **Tobacco Rattle**, CA.

WHITE RUST. **Albugo occidentalis**, AR, OK, TX.

WILT. **Verticillium** sp., NY.

Downy mildew is the outstanding spinach disease, with all varieties susceptible in some degree because of physiologic races, although Dixie Market, Savoy Supreme, and others are considered resistant. To reduce blight, use virus-tolerant varieties.

SPONDIAS (Mombin)

NEMATODE, Root Knot. **Meloidogyne** sp., FL.

SPOT ANTHRACNOSE; Mombin Scab. **Sphaceloma spondiadis**, FL.

SQUASH and PUMPKIN (*Cucurbita*)

ANTHRACNOSE. **Colletotrichum lagenarium**, CT to NJ, KS, TX.

BACTERIAL, MLO. **Aster Yellows**.

BACTERIAL Soft Rot. **Erwinia carotovora** pv. **carotovora**, cosmopolitan; **E. aroideae**.

BACTERIAL Spot. **Xanthomonas cucurbitae**, CT, GA, IL, IN, MD, MA, MI, WA, WI; **Pseudomonas syringae** pv. **lachrymans**; **P. syringae** pv. **syringae**, GA.

BACTERIAL Wilt. **Erwinia tracheiphila**, NM.

BLIGHT, Blossom. **Choanephora cucurbitarum**, brown rot of fruit, ME to FL, MI, OK, TX.

BLIGHT, Gummy Stem; Black Rot of Fruit. **Mycosphaerella citrullina** (*M. melonis*), CT, FL, GA, MA, MI, NJ, NY.

BLIGHT, Leaf. **Alternaria cucumerina**, DE, MN, NJ, NY, NC, UT; **Phytophthora capsici**, also stem rot, fruit rot, AR, CA, IL, NM, NC, TX.

BLIGHT, Southern. **Sclerotium rolfsii**, AL, FL, GA.

BLIGHT, Stem. **Fusarium tabacinum** (*Plectosporium tabecinum*), AL, IL, NY, TN.

DAMPING-OFF. **Pythium debaryanum**, CT, WI.

DOWNY MILDEW. **Pseudoperonospora cubensis**, ME to AL, CA, IA, LA, TX.

LEAF NECROSIS and TWIG CANKER. **Pseudomonas syringae**, **Pseudomonas marginolis**, Quebec, Canada.

LEAF SPOT. **Ascochyta** sp.; **Cercospora cucurbitae**, AL, DE, NJ, WI; **C. citrullina**, AL; **Gloeosporium** sp., IL; **Phyllosticta cucurbitarum**, IN, NY; **P. orbicularis**, DE, PA, NY; **Septoria cucurbitacearum**, IL, MA, NY, WI; **Stemphylium cucurbitacearum**, IN, OH.

MOLD, Seed. **Alternaria tenuis**, cosmopolitan; **A. radicina**, occasional; **Curvularia trifolii**, CT.

NEMATODE, Root Knot. **Meloidogyne** spp., MD to FL, TX.

NEMATODE, Stubby Root. **Trichodorus christiei**.

NONPARASITIC. **Blossom-End Rot**. On summer squash when hot, dry weather follows a cool, rainy spell.

Chlorosis. Nutrient deficiency: manganese, interveinal chlorosis, nitrogen, leaf yellowing and chlorosis of bud end of fruit.

Bronzing. Marginal Necrosis; Fruit Deformity. Potassium deficiency.

POWDERY MILDEW. **Golovinomyces cichoracearum**, general; **Sphaerotheca fuliginea**, CA, IN, MI, NY, NC.

ROT, Blossom-end; Root, **Pythium aphanidermatum**, AZ, CA, MD.

ROT, Charcoal. **Macrophomina phaseolina**, OR, TX.

ROT, Fruit. **Acidiovorax avenae** (syn. **Pseudomonas pseudoalcaligenes** subsp. **citrulli**), GA; **Alternaria** sp., MA, NH, NJ, NY, OR, VT, WA; **Diplodia natalensis**, TX; **Coniosporium fairmani**, black mold; **Botrytis cinerea**, gray mold; **Fusarium** spp.; **F. solani**, AR, **F. solani** f. sp. **cucurbitae**, also root and stem rot, CA, CT, NY, OR, WA; **F. equiseti**, AR, CT; **F. acuminatum**, CT; **F. graminearum**,

CT; **F. avenaceum**, CT; **Phytophthora cactorum**, AZ; **P. nicotianae**, NC; **Pythium ultimum**, CA; **Rhizopus stolonifer**, general in storage after injury; **Sclerotinia sclerotiorum**, ID, MA, ME, NH, NY, WA; **Trichothecium roseum**, pink mold, MA; **Phoma subvelata**, also leaf spot, stem rot, TX.

ROT, Root. **Phymatotrichum omnivorum**, TX; **Rhizoctonia solani**, TX.

SCAB; Leaf Spot; Storage Rot. **Cladosporium cucumerinum**, CT, MD, MA, NJ, NY, OR, WA.

VIRUS. **Bromegrass Mosaic**. General, IA, SD on squash (*Cucurbita pepa*); **Lettuce Mosaic**, NY; **Squash Mosaic**, CA, SD; **Southern Squash Mosaic**; **Cucumber Mosaic**; **Beet Curly Top**; **Tobacco Ring Spot**; **Prunus Ring Spot**; **Watermelon Mosaic**; **Zucchini Yellow Mosaic**, AR, CA, CT, FL, LA, NJ, NY, OH, SC; **Squash Leaf Curl**, CA; **Lettuce Infectious Yellows**, AZ, CA, TX; **Cucurbit Leaf Curl**, AZ, FL, TX, Mexico; **Beet Pseudo Yellows**, CA; **Potato Leaf Roll**, Pacific Northwest; **Cucurbit Yellow Stunting**, AZ, Mexico.

WILT. **Fusarium oxysporum** f. sp. **niveum**, CA, IL, MI, WI; **Verticillium albo-atrum**, OR.

WILT, Anasa. Feeding injury from the squash bug.

In general, diseases and control measures are the same as for cucumbers. Some squash varieties are injured by sulfur dusts. Acorn and butternut squashes are resistant to bacterial wilt.

STRAWBERRY (*Fragaria*)

ANTHRACNOSE; Crown Rot. **Colletotrichum acutatum**, CT; **C. fragariae**, FL, LA; **Gloeosporium** sp., MD.

ANTHRACNOSE, Fruit. **Colletotrichum gloeosporioides**, FL, OH; **C. acutatum**, NY.

BACTERIAL Angular Leaf Spot. **Xanthomonas fragariae**, FL, KY, MN, WI.

BACTERIAL Blossom Blight. **Xanthomonas campestris** var. **fragariae**, CA; **Pseudomonas marginalis**, CA.

BACTERIAL, Phytoplasma. **Aster Yellows**, strawberry green petal; **Lethal Disease Decline** or **Peach Western X Disease**. WA, OR; **Clover Proliferation**, CA, FL, MD; **Clover Yellow Edge** and **STRAWB2**, MD.

BACTERIAL Soft Rot. **Erwinia carotovora**, MA.

BLIGHT, Blossom. **Cladosporium cladosporioides**, CA.

BLIGHT, Gray Mold. **Botrytis cinerea**, general on fruit.

BLIGHT, Leaf. **Dendrophoma obscurans**, angular leaf spot, MA, to FL, MI, MN, NE, OR, TX; **Phomopsis obscurans**, CT.

BLIGHT, Southern. **Sclerotium rolfsii**, AL, FL, NC, TX.

BROOMRAPE. **Orobanche** sp., WA.

DOWNY MILDEW. **Peronospora fragariae**, IA.

GALL, Leaf. **Synchytrium aureum**, on native spp., WI; **S. fragariae**, root gall, CA, WA.

LEAF BLOTCH. **Gnomonia fructicola**, also fruit rot; **G. comari**, CA.

LEAF SCORCH. **Diplocarpon earliana**, general.

LEAF SPOT. **Cercospora fragariae**, LA; **C. vexans**, CA, NY, WI; **Gloeosporium** sp., IL, MA, MI, NC, PA, UT, associated with black root; **Mycosphaerella**

fragariae, common leaf spot, black-seed disease, general; **M. louisianae**, purple leaf spot, LA, MS, NC.

NEMATODE, Dagger. **Xiphinema americanum**; **X. diversicaudatum**; **X. chambersi**.

NEMATODE, Leaf and Stem. **Ditylenchus dipsaci**, CA, ID, NC, OR, TX, WA.

NEMATODE, Lesion. **Pratylenchus penetrans**; **P. coffeae**; **P. pratensis**; **P. scribneri**.

NEMATODE, Root Knot. **Meloidogyne hapla**.

NEMATODE, Spring Dwarf. **Aphelenchoides fragariae**, MA to FL, AR, CA, MI, TX, TN, WA; Summer Dwarf, Crimp. **A. besseyi**, southeastern and Gulf states, also AR, CA, DE, IL, TX.

NEMATODE, Sting. **Belonolaimus gracilis**.

NEMATODE, Stubby Root. **Trichodorus christiei**.

NONPARASITIC. **Black Root, Brown Root**. Winter injury, defective drainage, soil toxins, widespread.

Chlorosis. Iron deficiency.

Variegation, June Yellows. Chlorophyll deficiency, especially in Blake more, Progressive, and related varieties, general.

POWDERY MILDEW. **Sphaerotheca macularis**, general but rare in South, reported in FL, CA.

ROT, Crown. **Phytophthora cactorum**, NY; **Sclerotinia sclerotiorum**, MD to FL, CA, IL, IA, MN, TX; **Sclerotium rolfsii**, AL, FL, NC, TX; **Macrophomina phaseolina**, CA, FL, Argentina.

ROT, Fruit. **Penicillium** spp., secondary; **Pezizella lythri**, also leaf spot, root rot, ME to FL, AK, IL, IA, MI, OH, OK, OR, WI; **Phytophthora cactorum**, leather rot, AL, AR, IL, KY, LA, MD, MI, MO, OH, TN, VA; **Rhizopus nigricans**, black rot, cosmopolitan; **Rhizoctonia solani**, hard brown rot, widespread; **Sphaeropsis** sp., CA, CO, IL, IA, MN.

ROT, Gray Mold. **Botrytis cinerea**, SC.

ROT, Red Stele; Brown Core Root Rot. **Phytophthora fragariae**, ME to VA, IA, OK; also CO, MI, WI, Pacific states.

ROT, Root. **Armillaria mellea**, CA, OR, WA; **Cylindrocladium scoparium**, OR, TN; associated with black root; **Leptosphaeria coniothyrium**; **Fusarium** spp.; **Olpidium brassicae**, WA; **Pythium** spp., widespread; **Ramularia** spp., ID, OR, WA; **Idriella lunata**, CA, MD; **Phomopsis obscurans**, OH; **Macrophomina phaseolina**, FL, Argentina.

ROT, Root, Stolon and Petiole. **Pestalotia longisetula**, IL, OR.

SLIME MOLD. **Diachea leucopodia**, CA, IL, KS, LA, MS, MO, TN, TX; **Fuligo septica**, IA, KS, MN, NE, NJ, TX, WA; **Mucilago spongiosa**, KS, MN, MO, NE, OK; **Physarum cinereum**, IL, NE, NJ.

SOOTY MOLD. **Scoria spongiosa**, ME.

VIRUS. **Strawberry Crinkle**, CA, ID, MN, NY, OR, WA; **Strawberry Yellow Edge**, Xanthosis, CA, NC, OR, WA, and other states introduced from West Coast; **Strawberry Leaf Roll**, MD, NJ, NY, OR, VT; **Strawberry Mottle**; **Strawberry Multiplier Disease**, WI; **Strawberry Necrotic Shock**, CA; **Strawberry Pallidosis**, CA, OR, Mid-Atlantic, British Columbia, Canada; **Strawberry Pseudo**

Mild Yellow Edge; **Strawberry Mild Yellow Edge**, CA; **Strawberry Vein Necrosis**, MN; **Strawberry Vein Banding**; **Strawberry Witches' Broom**, ID, MN, MT, NY, OR, WA; **Tomato Ringspot,** PA; **Apple Mosaic**, OR; **Chlorotic Peacock Pattern**, OR; **Cucurbit Yellows**, OR; **Strawberry Phyllody**, MD, WV; **Tobacco Streak**, MD.

WILT. **Verticillium albo-atrum**, CA, FL, NY, OR, WA; **Colletotrichum acutatus**, AR, CA, FL, MS, MO; **Fusarium oxysporum**, CA.

Red Stele is, except in the deep South, of first importance on strawberries. The roots rot, and above-ground parts are stunted and wilted. Choose resistant varieties, such as Stelemaster, Sparkle, Surecrop, Midway, and Fairland. Purchase plants certified free from nematodes and virus diseases. Mulching helps to prevent fruit rots.

U.S. Department of Agriculture Farmers' Bulletin 1891 gives an excellent discussion of "Diseases of Strawberries". The American Phytopathological Society, St. Paul, MN has published a Compendium of Strawberry Diseases.

SUGAR BEET (Beta vulgaris)

POWDERY MILDEW. **Erysiphe polyoni**, ND.

ROT, Stalk and Blight. **Fusarium solani**, CA, Pacific Coast.

YELLOWS. **Fusarium Yellows**; **Fusarium oxysporum**, MI; **F. sulphureum**, MN.

SWEET POTATO (*Ipomoea batatas*)

BACTERIAL Soft Rot. **Erwinia carotovora**, CT, SC.

BLIGHT, Leaf. **Phyllosticta batatas**, occasional NJ to FL, TX, KS, usually in the South; **Choanephora cucurbitarum**, FL, leaf mold.

BLIGHT, Southern; Cottony Rot. **Sclerotium rolfsii**, general in South.

CANKER, Stem. **Fusarium solani**, NC. LEAF SPOT. **Alternaria** sp., secondary, occasional to general; **Cercospora** sp., FL, OK; **Septoria batat icola**, Occasional NJ to AL, IA, TX.

LEAF SPOT, Blight. **Acremonium crotocinigenum** (= *Cephalosporium cinnamomeum*), CA, FL.

NEMATODE, Lance. **Hoplolaimus coronatus**; Lesion, **Pratylenchus** sp.

NEMATODE, Root Knot. **Meloidogyne incognita** and **M. incognita-acrita**.

NEMATODE, Stem, Bulb. **Ditylenchus dipsaci**, causing brown ring, NJ.

NEMATODE, Sting. **Belonolaimus gracilis**.

NEMATODE, Stunt. **Tylenchorhynchus claytoni**.

NONPARASITIC. **Internal Breakdown**. Perhaps from chilling.

Internal Brown Spot. Boron deficiency.

Intumescence. High water intake and retarded transpiration.

ROT, Black. **Ceratocystis fimbriata**, general on roots, stems; **Diplodia theobromae** (*Physalospora rhodina*), Java black rot, general in South.

ROT, Charcoal. **Macrophomina phaseoli**, NJ to FL, TX, CA, KS.

ROT, Gray Mold. **Botrytis cinerea**, on sprouts, cosmopolitan.

ROT, Root. **Diaporthe batatis**, dry rot, stem rot, NJ to FL, MO, TX; **Helicobasidium purpureum**, KS, TX; **Pyrenochaeta terrestris**, pink root, CA; **Phymatotrichum omnivorum**, AZ, NM, OK, TX; **Phytophthora** sp., NJ, PA; **Plenodomus destruens**,

foot and storage rot, NJ to FL, CA, IA, LA; **Pythium** spp., mottle necrosis, leak, ring rot; **Fusarium solani**, NC; **Geotrichum candidum**, LA, NC.
ROT, Soil; Pox. **Streptomyces ipomoea**, general, NJ to FL, TX, AZ, CA, IA.
ROT, Sprout; Stem Canker. **Rhizoctonia solani**, occasional, NJ to FL, AZ, CA, OH, TX, WA.
ROT, Storage. **Fusarium** spp., general; **Mucor racemosus**, occasional after chilling; **Penicillium** sp., blue mold, cosmopolitan; **Rhizopus** spp., soft rot, general; **Sclerotinia** sp., NC; **Trichoderma** spp., DE, NJ; **Aspergillus** spp., secondary.
RUST. **Coleosporium ipomoeae** (II, III), AL, MS; 0, I on pine.
SCURF. **Monilochaetes infuscans**, general.
SLIME MOLD. **Fuligo violacea**, AL, DE, NJ, TX; **Physarum cinereum**; **P. plumbeum**, AL, AR, KY, MS, NC, VA.
VIRUS. **Sweet Potato Feathery Mottle** (Yellow Dwarf); **Sweet Potato Internal Cork**, GA, LA, MD, MS, NC, SC, TN, VA; **Sweet Potato Mosaic** (Tobacco Mosaic); **Tobacco Ring Spot**; **Beet Curly Top**; **Sweet Potato Russet Crack; Sweet Potato Chlorotic Stunt,** NC.
WHITE RUST. **Albugo ipomoeae-panduratae**, general.
WILT. **Fusarium oxysporum** f. sp. **batatas**, stem rot, general; **Verticillium alboatrum**, CA. Select varieties resistant to Fusarium stem rot, in general the Spanish group. Heat treatment eliminates some viruses. Hot water (10 min at 120°F) controls scurf. Variety Nemagold is resistant to root-knot nematodes.

SWISS CHARD (*Beta vulgaris* var. *cicla*)
BACTERIAL Leaf Spot. **Pseudomonas syringae** pv. **aptata**, CA.
BLIGHT, Phytophthora. **Phytophthora capsici**, IL.
BLIGHT, Southern. **Sclerotium rolfsii**, LA, SC.
DAMPING-OFF, Root Rot. **Rhizoctonia solani**, NY; **Pythium aphanidermatum**, CA.
DOWNY MILDEW. **Peronospora schachtii**, CA.
LEAF SPOT. **Cercospora beticola**, general; **Ramularia beticola**, WA.
MOLD, Seed. **Alternaria tenuis**, CA; **Stemphylium botryosum**, CA, WA.
NEMATODE, Root Knot. **Meloidogyne** sp., LA.
NONPARASITIC. **Heart Rot**; Cracked Stem. Boron deficiency, NY, WA.
ROT, Crown. **Sclerotinia** sp., MS.
ROT, Root. **Phymatotrichum omnivorum**, TX.
RUST. **Uromyces betae** (II), CA, OR.
VIRUS. **Beet Curly Top**, AZ, CA, OR; **Beet Mosaic**, AZ, CA, WA; **Yellow Net**, unidentified, CA.

T

THISTLE, IVORY AND MILK (*Silybum*)

Rust. **Puccinia punctiformis**, MD.

Smut, Biological Control. **Microbotryum silybum**, MD.

TOMATILLO (*Physalis*)

Smut, Leaf. **Entyloma austral**.

Virus. **Tomato Yellow Leaf Curl**, Guatemala, Sinaloa, Mexico; **Tomato Spotted Wilt**, CA, Mexico.

TOMATO (*Lycopersicon*)

Anthracnose. **Glomerella phomoides** (*Colletotrichum coccodes*), chiefly ripe rot of fruit, sometimes on leaves, general, especially in Northeast; **Colletotrichum dematium** var. **truncata**, MS; **C. graminicola**, MS; **C. trichellum**, MS; **C. glycines**, MS; **C. gossypii**, MS.

Bacterial Blight. **Pseudomonas viridiflava**, FL; **P. huttiensis**, FL.

Bacterial Canker. **Clavibacter michiganense**, birds-eye spot, general, most frequent in North and West, Baja, Mexico.

Bacterial Crown Gall. **Agrobacterium tumefaciens**; in experiments, **A. rhizogenes**, hairy root.

Bacterial, MLO. **Aster Yellows**; **Tomato Big Bud**, NY.

Bacterial, Pith Rot. **Erwinia chrysanthemi**, MA; **Pseudomonas corrugata**, CA, FL, LA, MA; also Stem necrosis, FL, Mexico.

Bacterial, Seedling Blight. **Bacillus polymyxa**, NY.

Bacterial Soft Rot. **Erwinia aroideae** and **E. carotovora**, cosmopolitan.

Bacterial Speck. **Pseudomonas syringae** pv. **tomato**, occasional East and central states, also CA, OK, TX, VA; **P. syringae** pv. **tomato** race 1, CA.

Bacterial Spot. **Xanthomonas vesicatoria**, on fruit, leaves, sometimes stem cankers, northeastern, central, and southern states; **X. campestris** pv. **vitians**, OH; **X. gardneri**, MI, OH, PA.

Bacterial Stem Rot. **Erwinia chrysanthemi**, FL.

Bacterial Wildfire. **Pseudomonas syringae** pv. **tabaci**, WI.

Bacterial Wilt. **Pseudomonas solanacearum**, general, MA to IL and southward.

R.K. Horst, *Field Manual of Diseases on Fruits and Vegetables*,

DOI 10.1007/978-94-007-5974-9_21,

BLIGHT, Aerial. **Pythium myriotylum**, FL.
BLIGHT, Blossom. **Sclerotinia** sp., CA, FL, NY, OH.
BLIGHT, Early; Collar Rot; Fruit Rot. **Alternaria solani**, general.
BLIGHT, Foliar. **Rhizoctonia solani**, NC.
BLIGHT, Gray Mold/Stem Canker. **Botrytis cinerea**, CA, Mexico.
BLIGHT, Late; Fruit Rot. **Phytophthora infestans**, general in humid regions and seasons, especially East and Southeast, AK.
BLIGHT, Southern. **Sclerotium rolfsii**, VA to FL and TX, CA, occasional in North.
BROOMRAPE. **Orobanche ludoviciana**, WY; **O. racemosa**, CA, NJ, NY.
CANKER, Stem. **Helminthosporium** sp., TX; **Myrothecium roridum**, also fruit rot, OH, TX, VA, WI.
DAMPING-OFF. **Rhizoctonia solani**, also collar rot, stem canker, cosmopolitan; **Pythium** spp., cosmopolitan.
DODDER. **Cuscuta** spp., CA, ID, MD, NY, TX.
DOWNY MILDEW. **Peronospora tabacina**, GA, NC, SC.
FRUIT SPOT. **Pullularia pullulans**, WV.
LEAF SPOT. **Ascochyta lycopersici**, DE, FL, NJ, NC, OR, VA, WI; **Septoria lycopersici**, general except Northwest; **Phyllosticta hortorum**, NY; **Cercospora** spp.; **Stemphylium solani**, gray leaf spot, FL, GA, IN, LA, NJ, NC, SC, TN, TX, VA; **S. floridanum**; **Cladosporium oxysporum**, NY; **Cristulariella moricola**, NY.
LEAF SPOT, Target, fruit necrotic pit, freckles. **Corynespora cassiicola**, FL.
MOLD, Blue. **Penicillium oxalicum**, Mexico.
MOLD, Leaf. **Botryosporium pulchrum**, PA, TX; **Chaetomium bostrychodes**, TX; **Cladosporium fulvum**, general under glass, occasional outdoors, IA.
NEMATODE, Awl. **Dolichodorus heterocephalus**.
NEMATODE, Dagger. **Xiphinema americanum**.
NEMATODE, Golden. **Heterodera rostochiensis** (*Globodera rostochiensis*); tobacco cyst, **H. tabacum**; soybean cyst, **H. glycines**, IA, MN, OH.
NEMATODE, Lance. **Hoplolaimus coronatus**.
NEMATODE, Lesion. **Pratylenchus pratensis**, MD, NJ.
NEMATODE, Reniform. **Rotylenchulus reniformis**.
NEMATODE, Root Knot. **Meloidogyne arenaria** var. **thamesi**; **M. hapla**, Pacific Northwest; **M. incognita**; **M. javanica**; **M. chitwoodi**, Pacific Northwest; **M. floridensis**, FL.
NEMATODE, Spiral. **Helicotylenchus nannus**.
NONPARASITIC. **Blossom-End Rot**. Unbalanced moisture and calcium deficiency, general.
Blotchy Ripening. Malnutrition, potassium deficiency.
Catface. Fruit abnormalities from growth disturbances.
Cloudy Spot. Feeding punctures of plant bugs.
Cuticle Crack. Of green fruit; high soil moisture and air temperature.
Fasciation. Genetic abnormality. Unbalanced nutrition.
Leafroll. Excessive soil moisture with starch congestion in leaves.
Oedema. Leaf hypertrophy from excessive water absorption and reduced transpiration.

Pockets; **Puffing**. Fruit defect from factors adversely affecting pollination and growth.

Psyllid Yellows. Toxemia from psyllid feeding, western states; **Candidatus Liberibacter psyllaurous**, AZ, CO.

Russeting. Mite injury.

Sunscald. Fruit injury in heat on plants defoliated by disease.

Walnut Wilt. Toxemia from root excretions of walnut trees.

PHYTOPLASMA, Citrus Huanglongbing. **Candidatus Liberibacter asiaticus**, FL.

PHYTOPLASMA, Leaf Cup Stunt. **Candidatus Liberibacter solanacearum**, New Zealand.

PHYTOPLASMA, Little Leaf. **Phytoplasma asteris**, Mexico.

PHYTOPLASMA, Yellowing Disease. **Candidatus Liberibacter solanacearum**, TX, Central America, Mexico, Baja, CA.

POWDERY MILDEW. **Erysiphe polygoni**, on seedlings indoors, NC; **Erysiphe** sp., CT, FL, NJ, NY, Canada; **Leveillula taurica**, AZ, CA, UT; **Oidiopsis taurica**, CA; **Oidium** sp., CA; **O. neolycopersici**, FL, MI, Mexico.

ROT, Black Root. **Thielaviopsis basicola**, CA.

ROT, Brown, Root. **Pyrenochaeta lycopersici**, MA.

ROT, Buckeye; Stem Rot. **Phytophthora parasitica**, MA to FL, AZ, CA, IL, TX; **P. cactorum**, NY, PA, WI; **P. capsici** United States, Mexico; **P. cryptogea**; **P. drechsleri; P. nicotianae,** NM.

ROT, Charcoal. **Macrophomina phaseoli**, CA, TX.

ROT, Fruit. **Alternaria** sp., black mold; **Aspergillus** spp., green and yellow mold; **Alternaria alternata**, NY; **Brachysporium tomato**, KS, TX; **Cladosporium herbarum**, green mold; **Diaporthe phaseolorum**, MS, TX; **Diplodia theobromae**, AL; **Glomerella cingulata**, FL, LA, ME, MI, NJ, NY; **Isaria clonostachoides**, VA; **Mucor mucedo**, MD; **Nematospora coryli**, cloudy spot, CA, FL, GA; **Nigrospora oryzae**, CA; **Oospora lactis**, sour rot, cosmopolitan; **Phoma destructiva**, black spot, nearly general; **Pleospora lycopersici**, CA; **Rhizopus stolonifer**, cosmopolitan in transit; **Sclerotinia minor**, TX; **Sporotrichum** sp., IN, TX; **Trichothecium roseum**, MD, NC, OH, Brazil; **Gibberella zeae**; **Phoma** sp.; **Myrothecium** sp.; **Eremothecium coryli**, CA.

ROT, Gray Mold; Ghost Spot. **Botrytis cinerea**, occasional on foliage, fruit, stems.

ROT, Nailhead Spot. **Alternaria tomato**, on fruit, stems, CT to FL, ND, TX.

ROT, Root. **Aphanomyces cladogamus**; **Colletotrichum atramentarium**, black dot; **Pyrenochaeta terrestris**, secondary, IA, IL, NJ; **Phymatotrichum omnivorum**, AZ, OK, TX; **Plectospira myriandra**, VA; **Thielaviopsis basicola**, TX; **Fusarium oxysporum**, FL, NE, NH, OH, (different from wilt); **F. oxysporum** f. sp. **radicis**, crown, FL, NH, OH, PA, TX. Also **F. oxysporum** f. sp. **radicis-lycopersici**, TX; **F. solani** f. sp. **eumartii**, CA.

ROT, Root (and stunting of seedlings). **Pythium myriotylum** and **P. arrhenomanes**, FL.

ROT, Root and Wilt. **Colletotrichum coccodes**, CA.

Rot, Stem, Fruit. **Sclerotinia sclerotiorum**, occasional in greenhouses and in South; **Pythium myriotylum**.

Scab, Powdery. **Spongospora subterranea**, PA.

Viroid. **Mexican Papita**. British Columbia, Canada; **Tomato Chlorotic Dwarf**, AZ, Mexico; **Potato Spindle Tuber**, CA, MD.

Virus. **Tobacco Mosaic**, general, strains causing fernleaf and internal browning of fruit; **Cucumber Mosaic**, also causing shoestring; **Beet Curly Top**; **Western Yellow Blight**; **Tomato Aspermy**, AZ; **Tomato Ring Mosaic**; **Tomato Ring Spot,** FL (South); **Tomato Spotted Wilt**, CA, OK, TN, Mexico; **Tomato Streak** (Tobacco Mosaic plus Potato Virus X); **Tomato Yellow Net**; **Tomato Yellow Top**; **Witches' Broom**; **Tobacco Etch**, may cause severe mosaic; **Rugose Mosaic** (due to Potato Mottle Virus and Potato Virus Y); **Pseudo Curly Top**, FL; **Tobacco Streak**, CA; **Tomato Mosaic**, CA; **Tomato Mottle**, FL, SC, TN, VA; **Tomato Infectious Chlorosis**, CA, GA, Costa Rica; **Tomato Bushy Stunt**, CA; **Tomato Black Ring**; **Tomato Pseudo Curly Top**, FL; **Tomato Top Necrosis**, IL, IN, MO; **Tomato Yellow Leaf Curl** CA, FL, GA, HI, KY, LA, MS, NC, SC, TX, Guatemala, Sinaloa, Mexico; **Serrano Golden Mosaic**, AZ; **Potato Leafroll**, Pacific Northwest; **Pepper Golden Mosaic**, CA, Mexico; **Pepino Mosaic**, AZ, CO, TX, Ontario Canada, Mexico; **Tomato Chino La Paz**, Baja CA, Mexico; **Tomato Severe Leaf Curl**, Mexico; **Chocolate Spot** (**Tomato Torrado Virus**), Guatemala, Panama; **Necrosis Disease**, CA; **Candidatus Liberibacter solanacearum**, NM, Mexico; **Pelargonium Zonate Spot**, CA; **Tomato Golden Mottle,** Mexico; **Tomato Apex Necrosis**, Mexico.

Wart. **Synchytrium endobioticum**, PA (see under Galls).

Wilt. **Fusarium oxysporum** f. sp. **lycopersici**, general, Mexico, Race 3, FL, TN; **Verticillium alboatrum**, occasional in all regions.

Choose tomato varieties resistant to Fusarium wilt; purchase seed certified free from virus disease. In setting seedlings discard any showing galls of root-knot nematodes. Do not smoke while handling, and do not use tobacco stems as a mulch, although other mulches are helpful. Spray for late blight and other foliage diseases. State and federal agencies give warning when late blight is imminent and it is time to start spraying. See USDA Agricultural Handbook 203 for an excellent discussion of tomato diseases with fine illustrations.

TURNIP (*Brassica rapa*)

Anthracnose. **Colletotrichum higginsianum**, NY to FL, TX.

Bacterial Black Rot. **Xanthomonas campestris**, ME to FL, MN, TX.

Bacterial Blight. **Pseudomonas syringae** pv. **alisalensis,** CA.

Bacterial Crown Gall. **Agrobacterium tumefaciens**, KS.

Bacterial Leaf Spot. **Pseudomonas syringae** pv. **maculicola**, OH; **Xanthomonas campestris** pv. **campestris**, SC.

Bacterial Soft Rot. **Erwinia carotovora**, general in field, transit, storage.

Bacterial Spot. **Pseudomonas maculicola**, CT, GA, MA, TX; **Xanthomonas vesicatoria** pv. **raphani**, IN.

Blackleg. **Phoma lingam**, CT, MA, WA.

BLIGHT, Southern. **Sclerotium rolfsii**, FL, TX.

CLUB ROOT. **Plasmodiophora brassicicola**, ME to NC, CA, CO, MN, TX, WA.

DAMPING-OFF. **Pythium ultimum**, WI; **Rhizoctonia solani**, also root, stem and storage rot, general.

DOWNY MILDEW. **Peronospora parasitica**, MA to FL, IL, TX.

LEAF SPOT. **Alternaria brassicae**, gray leaf spot, general; **A. oleracea**, black leaf spot, CT, FL, MD, MA, NJ, NC, TX; **Alternaria raphani**, AZ; **Cercospora brassicicola**, AL, FL, GA, MS; **Mycosphaerella brassicicola**, ring spot, OR; **Phyllosticta** sp., TX, WV; **Ramularia** sp., AL, FL, WA; **Septomyxa affine**, AL.

LEAF SPOT; White Spot. **Cercosporella brassicae**, MA to FL, IN, OR, TX; **Pseudocercosporella capsellae**, CA.

MOLD, Seed. **Alternaria tenuis**, cosmopolitan; **Stemphylium botryosum**, occasional; **Curvularia inaequalis**.

NEMATODE, Root. **Pratylenchus pratensis**, MD.

NEMATODE, Root Knot. **Meloidogyne** spp., AL, AZ, FL, OR, TX.

NONPARASITIC. **Brown Heart**. Boron deficiency, CA, MA, MN, VA, WI.

POWDERY MILDEW. **Erysiphe polygoni**, northeastern states to CA, FL, TX, WA; **Erysiphe cruciferarum**, WY.

ROT, Root. **Phymatotrichum omnivorum**, TX; **Pythium** sp., NY.

ROT, Watery Soft. **Sclerotinia sclerotiorum**, CT, MD, MS, TX, WA.

SCAB. **Streptomyces scabies**, CT, MI, NJ, Canada.

SCAB, Acid. **Streptomyces acidiscabies**, ME.

VIRUS. **Turnip Mosaic**, AL; **Beet Curly Top**, CA; **Radish Mosaic; Turnip Ringspot,** OH.

WHITE RUST; White Blister. **Albugo candida**, general.

WILT; Yellows. **Fusarium oxysporum** f. sp. **conglutinans**, IN, MS, TX.

The control of turnip diseases is about the same as for cabbage and other crucifers.

V

VALERIANELLA (Corn-Salad or Lamb's Lettuce)
DOWNY MILDEW. **Peronospora valerianellae**, CA.
GALL, Leaf. **Synchytrium aureum**, MS.
LEAF SPOT. **Septoria valerianellae**, MS, TX.
POWDERY MILDEW. **Golovinomyces orontii**, CA.
ROT, Root. **Phymatotrichum omnivorum**, TX; **Thielaviopsis basicola**, PA.
ROT, Stem and Crown. **Sclerotinia minor**, CA.

VIDALIA SWEET ONION (*Allium*)
DOWNY MILDEW. **Peronospora destructor**, GA.
ROT, Bulb. **Botrytis tulipae**, GA.

R.K. Horst, *Field Manual of Diseases on Fruits and Vegetables*,
DOI 10.1007/978-94-007-5974-9_22,

W

WALNUT (*Juglans*) (Includes Butternut, Black, English, and Japanese Walnuts)

ANTHRACNOSE; Leaf Spot. **Gnomonia leptostyla** (*Marssonina juglandis*), general.

BACTERIAL Blight. **Xanthomonas juglandis**, NY to GA, TX, and Pacific Coast, especially on English (Persian) walnut.

BACTERIAL Canker. **Erwinia nigrifluens**, CA; **E. rubrifaciens**, phloem canker, CA; **Sirococcus clavigignenti-juglandacearum**, NC.

BACTERIAL Crown Gall. **Agrobacterium tumefaciens**, occasional.

BLIGHT, Leaf. **Cylindrosporium juglandis**, AL and TN to TX, CA.

BLIGHT, Seedling. **Sclerotium rolfsii**, TX.

CANKER. **Sirococcus clavigignenti-juglandacearum to Ophiognomonia clavigignenti-juglandacearum**, Ontario, Canada; **Phomopsis oblonga**, CT.

CANKER; Branch Wilt. **Hendersonula toruloidea**, connected with sunburn.

CANKER; DIEBACK. **Diplodia juglandis**, widespread; **Dothiorella gregaria**, CA; **Exosporina fawcetti**, wilt, CA; **Melanconis juglandis**, widespread, especially in East; **Nectria** spp., widespread; **N. galligena**; **Cytospora** sp., AZ.

CANKER; Felt Fungus. **Septobasidium curtisii**, NC.

CANKER, Stem. **Fusarium solani**, KS; **Phytophthora citricola, NC**.

CANKER DISEASE. **Geosmithia**, AZ, CA, CO, ID, OR, UT, WA.

DIEBACK, Twig and Branch. **Neofusicoccum mediterraneum**, CA.

LEAF SPOT. **Cercospora juglandis**, KS, MA; **Phloeospora multimaculans**, TX; **Marssonina californica**, CA; **Ascochyta juglandis**, ring spot; **Grovesinia pyramidalis**, IL, IA, MN, OH, WV; **Mycosphaerella juglandis**, IL, IN, IA, NC; **Cristulariella pyramidalis**, IL. Also on *Bidens frondosa* (beggar-ticks), *Campsis radicans* (trumpet vine), *Chenopodium ambrosioides* (Mexican tea), *Commelina diffusa* (dayflower), *Cuphea petiolota* (blue waxweed), *Desmodium canescens* (tick clover), *Eupatorium coelestinum* (mist-flower), *E. rugosum* (white snakeroot), *Ipomoea hederscea* (morning-glory), *I. lacunosa* (morning-glory), *Lobelia inflata* (Indian tobacco), *L. siphilitica* (blue cardinal-flower), *Perilla frutescens* (beefsteak plant), *Phytolacca americana* (poke), *Platanus occidentalis* (sycamore), *Polygonum pennsylvanicum* (smart weed), *P. scandens* (false buckwheat), *Rumex*

DOI 10.1007/978-94-007-5974-9_23,

crispus (yellow dock), *Sida spinosa* (prickly mallow), *Solidago canadensis* (goldenrod), *Vitis palmata* (Catbird grape).

LEAF SPOT; Downy Spot; White Mold. **Microstroma juglandis**, wide-spread; **M. brachysporum**, general; **Alternaria arborescens**, **A. alternata**, **A. tenuissima** (leaf spot and brown/gray apical necrosis), AZ.

MISTLETOE. **Phoradendron serotinum (flavescens)**, IN southward, AZ, CA, NM.

MOLD, Nut. **Alternaria** sp., CA.

NEMATODE, Lesion. **Pratylenchus musicola**, CA; **P. vulnus**.

NEMATODE, Root Knot. **Meloidogyne** spp., TX.

NONPARASITIC. **Black End**, of Nuts. Probably drought injury, CA, OR.

Black Line, Girdle. Graft incompatibility.

Dieback. Boron deficiency, OR.

Erinose. Leaf galls from blister mites.

Leaf Scorch, **Sunscald**. NJ. OR, WA.

Little Leaf. Zinc deficiency, CA. Probably also causes Rosette.

Shrivel, Witches' Broom, Yellows. Cause unknown.

PHYTOPLASMA, **Witches' Broom**, GA, IL, IN, IA, OH.

POWDERY MILDEW. **Phyllactinia corylea**, IN, OH, OR; **Microsphaera alni**, wide-spread; **Erysiphe polygoni**, CA.

ROT, Collar. **Phytophthora cactorum**, CA; **P. cinnamomi**, MD to AL, LA.

ROT, Heart. **Fomes igniarius** and **F. everhartii**, widespread; **Polyporus sulphureus**, widespread.

ROT, Root. **Armillaria mellea**, cosmopolitan; **Phymatotrichum omnivorum**, TX; **Cylindrocladium** sp., TN.

ROT, Wood. **Fomes conchatus**; **Polyporus** spp.; **Poria** spp.; **Schizophyllum commune**, cosmopolitan; **Daedalea confragosa**; **D. quercina**, widespread.

SCAB. **Cladosporium** sp., MD, MN.

VIRUS. **Brooming Disease**, DC, GA, MD, NY.

Bacterial blight is the most serious disease of walnuts on the Pacific Coast, requiring several sprays. A toxin, juglone, has been considered injurious to many shrubs growing in the vicinity of black walnut roots; the toxin also causes vascular wilt-like symptoms in eggplant and tomato in home gardens.

WATER-CRESS (*Nasturtium officinale*)

DAMPING-OFF. **Rhizoctonia solani**, TX.

LEAF SPOT. **Cercospora nasturtii**, CA, CT, FL, IN, NH, TX, WI.

ROT, Root. **Pythium debaryanum**, TX; **Sporgospora subterranea** f. sp. **nasturtii**, FL, PA; **Phytophthora cryptogea**, and stem, FL.

RUST. **Puccinia aristidae**, AZ, CO, TX.

VIRUS. **Western Aster Yellows Phytoplasma**, HI.

VIRUS, Mosaic and Stunting. **Turnip Virus**, Brazil.

WHITE RUST; White Blister. **Albugo candida**, MN.

WATERMELON (*Citrullus*)

ANTHRACNOSE. **Colletotrichum lagenarium**, general, Pacific Coast; **C. orbiculare**, CA; **Marssonina melonis**, NY.

BACTERIAL, Angular Leaf Spot. **Pseudomonas pseudoalcaligenes** subsp. **citrulli**, GA.

BACTERIAL Fruit Blotch. **Acidovorax avenae** (syn. **Pseudomonas pseudoalcaligenes**) subsp. **citrulli**, DE, IL, IN, OR, TX.

BACTERIAL Rind Necrosis. **Erwinia** sp., TX.

BACTERIAL Soft Rot. **Erwinia aroideae**, WV; wilt, **E. tracheiphila**, rare.

BACTERIAL Spot. **Pseudomonas lachrymans**, MI.

BACTERIAL Wilt. **Erwinia tracheiphila**, NM.

BLIGHT, Gummy Stem; Fruit Spot. **Didymella bryoniae**, CA; **Mycosphaerella citrullina**, also stem-end rot, leaf spot, MA to FL, AZ, MO.

BLIGHT, Southern. **Sclerotium rolfsii**, also fruit rot, NC to FL, TX.

CANKER, Stem. **Myrothecium roridum**, GA.

DAMPING-OFF. **Rhizoctonia solani**, soil rot, leaf blight; **Pythium** spp., also blossom-end rot, foot rot.

DOWNY MILDEW. **Pseudoperonospora cubensis**, occasional from MA to FL, TX, WI; also CA.

FRUIT SPOT; Speck. **Cribropeltis citrullina**, IL.

LEAF SPOT. **Alternaria cucumerina**, general except Pacific Coast; **A. alternate** f. sp. **cucurbitae,** MD. **Cercospora citrullina**, NJ to FL, OH, TX; **Corynespora cassiicola**; **Myrothecium roridum**, GA.

NEMATODE, Root knot. **Meloidogyne arenaria**; **M. incognita**; **M. javanica**.

NONPARASITIC. **Blossom-End Rot**. Hot dry weather after cool, moist days. **Internal Browning**. Drought and nutritional deficiencies.

POWDERY MILDEW. **Golovinomyces cichoracearum**, AZ, CA, FL, GA, NY, NC, TX, VA. May make pimples in young fruit; **Podosphera xanttii**, FL.

ROT. **Diplodia** spp., stem-end, gray; MD to FL, AZ, KS; **Fusarium scirpi**, occasional in market; **Helminthosporium** sp., TX; **Rhizopus** spp., mushy soft rot.

ROT, Charcoal. **Macrophomina phaseoli**, TX.

ROT, Root. **Acremonium** sp., CA; **Phymatotrichum omnivorum**, AZ, TX; **Thielaviopsis basicola**, OR, UT; **Monosporascus cannonballus**, CA, Brazil.

ROT, Stem and Fruit. **Phytophthora cactorum**, AZ; **P. capsici**, CO; **P. citrophthora**, CA; **Sclerotinia sclerotiorum**, NJ, TX.

SCAB; Leaf Mold. **Cladosporium cucumerinum**, MD, NE.

VIRUS. **Watermelon Mosaic**, NY to FL, AZ, CA, IA, MI, TX, WA; **Beet Curly Top**; **Tobacco Ring Spot**; **Watermelon Stunt** (strain of Squash Mosaic); **Zucchini Yellow Mosaic**, CA; **Watermelon Curly Mottle**, AZ; **Papaya Ring Spot,** LA; **Squash Leaf Curl**, TX; **Lettuce Infectious Yellows**, AZ, CA, TX; **Tomato Spotted Wilt**, GA; **Cucurbit Leaf Crumple**, CA; **Squash Vein Yellowing**, IN; **Cucurbit Yellow Stunting**, AZ, Mexico.

WILT. **Fusarium oxysporum** f. sp. **niveum**, GA, general and Race 2, DE, IN, MD; Race 3, MD; **Verticillium albo-atrum**, CA, NH, OR; **V. dahlia**, TX, **Rhizoctonia solani AG-7**, IN.

Fusarium wilt is probably the major disease and resistant varieties are available. For anthracnose and leaf spots treat seed before planting, and start spraying or dusting when vines start to run.

Y

YAM, CINNAMON-VINE (*Dioscorea*)

LEAF SPOT. **Cercospora dioscoreae**, IL, IA, MD, MI, PA, WI; **Colletotrichum dioscoreae**, IL; **Phyllosticta dioscoreae**, SC, VA, WV; **Ramularia dioscoreae**, WI.

NEMATODE, Root Knot. **Meloidogyne** sp., NC.

ROT, Root. **Phymatotrichum omnivorum**, TX.

R.K. Horst, *Field Manual of Diseases on Fruits and Vegetables*,
DOI 10.1007/978-94-007-5974-9_24, © Springer Science+Business Media Dordrecht 2013

Part IV

List of Land-Grant Institutions and Agricultural Experiment Stations in the United States

For help in diagnosing and controlling plant diseases contact your County Ag Comissioners (CA) or county extension agent, your state Diagnostic Lab, your state Department of Agriculture or your Cooperative Extension Specialist.

Alabama: Auburn University, Auburn, AL 36849. http://www.aces.edu/dept/plantdiagnosticlab/

Alabama Cooperative Extension Service http://www.aces.edu/plantlabbham/

Alaska: University of Alaska and Alaska Department of Agriculture http://www.dnr.state.ak.us/ag/index.htm

Arizona: University of Arizona, Tucson, AZ 85721. http://ag.arizona.edu/PLP/plpext/

Arkansas: University of Arkansas, Fayetteville, AR 72701. http://www.aragriculture.org/pestmanagement/diseases/clinic/default.asp

California: Department of Food and Agriculture, Plant Pest Diagnostic Center. (sgaimari@cdfa.ca.gov)

Colorado: Colorado State University, Fort Collins, CO 80523. http://www.coopext.colostate.edu/jeffco/hort/clinic.htm

Connecticut: University of Connecticut, Storrs, CT 06269. (http://pronewengland.org)

Delaware: University of Delaware, Newark, DE 19716. http://ag.udel.edu/extension/pdc/index.htm

District of Columbia: See Rutgers. http://www.rcs.rutgers.eon/services/

Florida: University of Florida, Gainesville, FL 32611. http://plantpath.ifas.ufl.edu/

Georgia: University of Georgia, Athens, GA 30602. http://www.plant.uga.edu/Extension/Clinics/PDC.htm

Hawaii: University of Hawaii, Honolulu, HI 96822. http://www2.ctahr.hawaii.edu/adsc/

Idaho: University of Idaho, Extension Service, Parma, ID 83660. http://www.uidaho.edu/ag/plantdisease/

Illinois: University of Illinois, Urbana, IL 61802. http://plantclinic.cropsci.uiuc.edu/

Indiana: Purdue University, West Lafayette, IN 47907. http://www.ppdl.purdue.edu

Iowa: Iowa State University, Ames, IA 50011. http://www.isuplantdiseaseclinic.org

R.K. Horst, *Field Manual of Diseases on Fruits and Vegetables*,
DOI 10.1007/978-94-007-5974-9_25, © Springer Science+Business Media Dordrecht 2013

Kansas: Kansas State University, Manhattan, KS 66506. http://www.plantpath.k-state.edu/DesktopDefault.aspx?tabid=49

Kentucky: University of Kentucky, Lexington, KY 40546. http://www.ca.uky.edu/agcollee/plantpathology/extension/pdd_lab.html

Louisiana: Louisiana State University, Baton Rouge, LA 70803. http://www.lsuagcenter.com/en/our_offices/departments/PlantPathology_Crops_Physiology/Plant_Disease_Clinic/

Maine: University of Maine, Orono, ME 04473. http://pmo.umext.main.edu

Maryland: University of Maryland, College Park, MD 20742. http://www.plantclinic.umd.edu/

Massachusetts: University of Massachusetts, Amherst, MA 01003. (umassextension.org)

Michigan: Michigan State University, East Lansing, MI 48824. http://www.pestid.msu.edu/

Minnesota: University of Minnesota, St. Paul, MN 55108. http://www.extension.umn.edu/distribution/cropsystems/DC3170.html

Mississippi: Mississippi Cooperative Extension Service, Mississippi State University, MS 39762. http://www.extensionplantclinics.msstate.edu/

Missouri: University of Missouri, Columbia, MO 65211. http://soilplantlab.missouri.edu/

Montana: Montana State University, Bozeman, MT 59717. http://scarab.msu.montana.edu/ipm/

Nebraska: University of Nebraska, Lincoln, NE 68583. http://plantpath.unl.edu/ppathdiagnostic.htm

Nevada: University of Nevada, or Nevada Department of Agriculture. http://www.nevada.gov/

New Hampshire: University of New Hampshire, Durham, NH 03824. http://ceinfo.unh.edu/Agric/AGPDTS/PlantH.htm

New Jersey: Rutgers, The State University, Milltown, NJ 08850. http://www.rce.rutgers.edu.services/

New Mexico: New Mexico State University, Las Cruces, NM 88003. http://plantclinic.nmsu.edu

New York: Cornell University, Ithaca, NY 14853. http://plantclinic.cornell.edu

North Carolina: North Carolina State University, Raleigh, NC 27695. http://www.ncsu.edu/pdic

North Dakota: North Dakota State University, Fargo, ND 58105. http://www.ag.ndsu.nodak.edu/diaglab/

Ohio: Ohio State University, Columbus, OH 43210. http://ppdc.osu.edu

Oklahoma: Oklahoma State University, Stillwater, OK 74078. http://entoplp.okstate.edu/Pddl/

Oregon: Oregon State University, Corvallis, OR 97331. http://web.science.oregonstate.edu/bpp/Plant_Clinic/index.htm

Pennsylvania: The Pennsylvania State University, University Park, PA 16802. http://www.ppath.cas.psu.edu/extension.html

Puerto Rico: Department of Agriculture, San Juan, PR. http://www.agricultura.gobierno.pr

Rhode Island: University of Rhode Island, Kingston, RI 02881. http://www.uri.edu/ce/ceec/plantclinic.html

South Carolina: Clemson University, Clemson, SC 29634. http://www.clemson.edu/agsrvlb

South Dakota: South Dakota State University, Brookings, SD 57007. http://plantsci.sdstate.edu/planthealth/subpage.cfm?ID=12

Tennessee: University of Tennessee, Nashville, TN 37211. http://soilplantandpest.utk.edu/

Texas: Texas A & M University, College Station, TX 77845. http://plantdiseaselab.tamu.edu/

Utah: Utah State University, Logan, UT 84322. http://utahpests.usu.edu

Vermont: University of Vermont, Burlington, VT 05405. http://pss.uvm.edu/pd/pdc/

Virgin Islands: Virgin Islands Extension Service, Kingshill, St. Croix, VI 00850. http://rps.uvi.edu/CES/index.html

Virginia: Virginia Polytechnic Institute, Blacksburg, VA 24061. http://www.ppws.vt.edu/~clinic/

Washington: WSU Research and Extension Center, Puyallup, WA 98371. http://www.puyallup.wsu.edu/plantclinic/index.html

West Virginia: West Virginia University Extension Service, Morgantown, WV 26506. http://www.mastergardener2008.com/

Wisconsin: University of Wisconsin, Madison, WI 53706. http://pddc.wisc.edu/, http://www.entomology.wisc.edu/entodiag.html

Wyoming: University of Wyoming Cooperative Extension Service, Laramie, WY 82071. http://ces.uwyo.edu/

Ontario: Pest Diagnostic Clinic, University of Guelph, Ontario, Canada, N1H 8J7. http://www.labservices.uoguelph.ca/units/pdc/

Glossary

Acervulus, pl. *Acervuli* A "little heap," an erumpent, cushionlike mass of hyphae bearing conidiophores and conidia, sometimes with setae; characteristic of the Melanconiales.

Acicular Needlelike.

Aeciospore Rust spore formed in an aecium.

Aecium, pl. *Aecia* A cluster-cup, or cuplike fruiting sorus in the rusts.

Aerobic Living or active only in the presence of oxygen.

Amoeboid Not having a cell wall and changing in form like an amoeba.

Allantoid Sausage-shaped.

Alternate Host One or other of the two unlike hosts of a heteroecious rust.

Anamorph State of life-cycle in which asexual spores, or none, are produced.

Annulus A ring; ringlike partial veil around stipe in the mushrooms.

Antheridium, pl. *Antheridia* Male sex organ in the fungi.

Anthracnose A disease with limited necrotic lesions, caused by a fungus producing nonsexual spores in acervuli.

Antibiotic Damaging to life; especially a substance produced by one microorganism to destroy others.

Apothecium, pl. *Apothecia* The cup-or saucer-like ascus-bearing fruiting body; in the Discomycete section of the Ascomycetes.

Appressorium A swelling on a fungus germ tube for attachment to host in early stage of infection; found especially in anthracnose fungi and rusts.

Ascocarp, or Ascoma Any structure producing asci, as an apothecium, perithecium.

Ascomycetes One of the three main groups of the fungi, bearing sexual spores in asci.

Ascospore Produced in ascus by free cell formation.

Ascus, pl. Asci Saclike, usually clavate cell containing ascospores, typically eight.

Aseptate Without cross-walls.

Asexual Vegetative, having no sex organs or sex spores; the imperfect stage of a fungus.

Autoecious Completing life cycle on one host; term used in rusts.

Bacteria Microscopic one-celled organisms increasing by fission.

Bactericide Substance causing death of bacteria.

R.K. Horst, *Field Manual of Diseases on Fruits and Vegetables*,
DOI 10.1007/978-94-007-5974-9,

Basidiomycetes Class 3 in the Fungi, characterized by septate mycelium, sometimes with clamp-connections, and sexual spores on basidia.

Basidiospore Spore produced on a basidium.

Basidium, pl. Basidia Club-shaped structure, which, after fusion of two nuclei, produces four basidiospores.

Binucleate Having two nuclei.

Blight A disease with sudden, severe leaf damage and often with general killing of flowers and stems.

Blotch A blot or spot, usually superficial.

Breaking, of a virus Loss of flower color in a variegated pattern, especially in tulips.

Canker A lesion on a stem; a plant disease with sharply limited necrosis of the cortical tissue.

Carrier Infected plant showing no marked symptoms but source of infection for other plants.

Catenulate In chains, or in an end-to-end series.

Cerebroid With brainlike convolutions or folds.

Chemotherapy Treatment of internal disease by chemical agents that have a toxic effect on the microorganism without injuring the plant.

Chlamydospore Thick-walled, asexual resting spore formed by the rounding up of any mycelial cell; also used for smut spores.

Chlorosis Yellowing of normally green tissue due to partial failure of chlorophyll to develop; often due to unavailability of iron.

Cilium, pl. Cilia Hairlike swimming organ on bacteria or zoospores.

Cirrhus, pl. Cirrhi A tendril or horn of forced-out spores.

Clamp-connections Outgrowths of hyphae which form bridges around septa, thus connecting two cells; in Basidiomycetes.

Clavate Club-shaped.

Cleistothecium, pl. Cleistothecia A perithecium without a special opening; in powdery mildews.

Coalesce Growing together into one body or spot.

Coenocytic Multinucleate; mycelia having no cell walls.

Columella Sterile central axis in a mature fruiting body.

Concentric One circle within another with a common center.

Conidiophore Simple or branched hyphae on which conidia are produced.

Conidium, pl. Conidia Any asexual spore except sporangiospore or chlamydospore.

Conk Term used in forestry for sporophores of Polyporaceae on trees.

Control Prevention of, or reduction of loss from, plant disease.

Coremium, pl. Coremia Synnema, a cluster of erect hyphae bearing conidia.

Coriaceous Like leather in texture.

Culturing Artificial propagation of organisms on nutrient media or living plants.

Cystidium, pl. Cystidia Sterile, often swollen cell projecting from hymenium in Basidiomycetes.

Damping-off Seed decay in soil, or seedling blight.

Decumbent Resting on substratum with ends turned up.

Decurrent Running down the stipe or stem.
Diagnosis Identification of nature and cause of a disease.
Dieback Progressive death of branches or shoots beginning at tips.
Defoliate To strip or become stripped of leaves.
Dichotomous Branching, frequently successive, into two more or less equal arms.
Dimidiate Having one half smaller than the other; of a perithecium, having outer wall covering only top half.
Discomycetes The cup fungi, a subclass of Ascomycetes; with apothecia.
Disease A condition in which use or structure of any part of the living organism is not normal.
Disinfection Freeing a diseased plant, organ, or tissue from infection.
Disinfestation Killing or inactivating disease organisms before they can cause infection; on surface of seed or plant part, or in soil.
Dissemination Transport of inoculum from a diseased to a healthy plant.
Disjunctor Cell or projection connecting spores of a chain.
Duster Apparatus for applying fungicides in dry form.

Echinulate Having small, pointed spines; used of spores.
Endoconidium, pl. Endoconidia Conidium formed within a hypha.
Enphytotic A plant disease causing constant damage from year to year.
Epiphytotic Sudden and destructive development of a plant disease over an extensive area, an epidemic.
Eradicant Fungicide One that destroys a fungus at its source.
Eradication Control of disease by eliminating the pathogen after it is already established.
Erumpent Breaking through surface of substratum.
Excentric Off center.
Exclusion Control of disease by preventing its introduction into disease-free areas.
Exudate Liquid discharge from diseased tissues.

Fasciation Joining side by side; a plant disease with flattened and sometimes curved shoots.
Fascicle A small bundle or cluster.
Filiform Threadlike.
Fimbriate Fringed, or toothed.
Flag A branch with dead leaves on an otherwise green tree.
Flagellum, pl. Flagella Whiplike organ on a motile cell; cilium.
Fruiting Body Fungus structure containing or bearing spores; mushroom, pycnidium, perithecium, apothecium, etc.
Fumigant A volatile disinfectant, destroying organisms by vapor.
Fungicide Chemical or physical agent that kills or inhibits fungi.
Fungi Imperfecti Fungi that have not been connected with the perfect or sexual stage; most are imperfect states of Ascomycetes.
Fungistatic An agent preventing development of fungi without killing them.
Fungus, pl. Fungi An organism with no chlorophyll, reproducing by sexual or asexual spores, usually with mycelium with well-marked nuclei.

Fusiform Spindle-like, narrowing toward the ends.
Fusoid Somewhat fusiform.

Gall Outgrowth or swelling, often more or less spherical,of unorganized plant cells as result of attack by bacteria, fungi, or other organisms.
Gametangium Gamete mother cell.
Gamete A sex cell, especially one formed in a gametangium.
Germ Tube Hypha produced by a germinated fungus spore.
Gill Lamella or hymenium-covered plate on underside of cap of a mushroom.
Girdle A canker that surrounds stem, completely cutting off water supply and thus causing death; girdling roots also cause death.
Glabrous Smooth.
Gleba Sporulating tissue in an angiocarpous fruit body.
Globose Almost spherical.
Gram-negative, Gram-positive Not being stained, and being stained, by the Gram stain used in classifying bacteria.

Haustorium, pl. Haustoria Special hyphal branch extended into living cell for purpose of absorbing food.
Heteroecious Undergoing different parasitic stages on two unlike hosts, as in the rusts.
Heterothallic Of a fungus, sexes separate in different mycelia.
Holocarpic Having all the thallus used for a fruiting body.
Homothallic Both sexes present in same mycelium.
Host Any plant attacked by a parasite.
Hyaline Colorless, or nearly transparent.
Hymenium Spore-bearing layer of a fungus fruiting body.
Hyperplastic Term applied to a disease producing an abnormally large number of cells.
Hypha, pl. Hyphae Single thread of a fungus mycelium.
Hypoplastic Term applied to a disease with subnormal cell production.
Hyphopodium, pl. Hyphopodia More or less lobed appendage to a hypha.
Hysterothecium Oblong or linear perithecium, sometimes considered an apothecium, opening by a cleft.

Immune Exempt from disease; having qualities that do not permit infection.
Immunization Process of increasing the resistance of a living organism.
Imperfect Fungus One lacking any sexual reproductive state.
Imperfect State State of life-cycle in which asexual spores, or none, are produced. See Anamorph.
Incubation Period Time between inoculation and development of symptoms that can be seen.
Indehiscent Of fruit bodies, not opening, or with no special method.
Infection Process of beginning or producing disease.
Infection Court Place where an infection may take place, as leaf, fruit, petal, etc.
Injury Result of transient operation of an adverse factor, as an insect bite, or action of a chemical.

Innate Bedded in, immersed.
Inoculation Placing of inoculum in infection court.
Inoculum Pathogen or its part, as spores, fragments of mycelium, etc., that can infect plants.
Inoperculate Not opening by a lid.
Intercellular Between cells.
Intracellular Within cells.
Intumescence Knoblike or pustulelike outgrowth of elongated cells on leaves, stems, etc., caused by environmental disturbances.

Lamella Gill.
Lesion Localized spot of diseased tissue.
Locule A cavity, especially one in a stroma.

Macroconidia Large conidia.
Macroscopic Large enough to be seen with the naked eye.
Medulla Loose layer of hyphae inside a thallus; body of a sclerotium.
Microconidia Very small spores, now considered spermatia of a fungus also having larger conidia.
Micron 1/1,000 mm, unit used for measuring spores.
Microscopic Too small to be seen except with the aid of a microscope; true of most of the fungus structures shown in line drawings in this book.
Mildew Plant disease in which the pathogen is a growth on the surface.
Molds Fungi with conspicuous mycelium or spore masses, often saprophytes.
Monoecious Male and female reproductive organs in same individual; in rusts, all stages of life cycle on single species of plant.
Multinucleate Several nuclei in same cell.
Mummy Dried, shriveled fruit, result of disease.
Muriform Having cross and longitudinal septa.
Mushroom An agaric fruit body.
Mycelium, pl. Mycelia Mass of fungus hyphae.
Mycelia Sterilia Fungi Imperfecti where spores, except for chlamydospores, are not present.
Mycoplasmalike Organism (MLO) A walless prokaryotic plant pathogen which has a single-unit membrane. See Phytoplasma.
Mycorrhiza, pl. Mycorrhizae Symbiotic, nonpathogenic association of fungi and roots.

Necrosis Death of plant cells, usually resulting in tissue turning dark.
Necrotic As an adjective, killing.
Nematicide Chemical or physical agent killing nematodes.
Nematodes Nemas, roundworms, eelworms, cause of some plant diseases.

Obligate Parasite A parasite that can develop only in living tissues, with no saprophytic stage.
Obovate Inversely ovate, narrowest at base.
Obtuse Rounded or blunted, greater than a right angle.

Oogonium, pl. *Oogonia* Female sex organ in the Oomycetes.
Oomycetes Subclass of the Phycomycetes, gametangia of unequal size.
Oospore Resting spore formed in a fertilized oogonium.
Operculate With a cover or lid, as in some asci.
Ostiole Porelike mouth or openings in papilla or neck of a perithecium or pycnidium.

Papilla, pl. *Papillae* Small, nipplelike projection.
Paraphysis, pl. *Paraphyses* A sterile hyphal element in the hymenium, especially in the Ascomycetes, usually clavate or filiform.
Paraphysoids Threads of hyphal tissue between asci, like delicate paraphyses but without free ends.
Parasite An organism that lives on or in a second organism, usually causing disease in the latter.
Pathogen Any organism or factor causing disease.
Pathogenic Capable of causing disease.
Pedicel Small stalk.
Perfect State Stage of life-cycle in which spores are formed after nuclear fission. See Teleomorph.
Peridium Wall or limiting membrane of a sporangium or other fruit body, or of a rust sorus.
Perithecium Subglobose or flasklike ascocarp of the Pyrenomycetes.
Phialide A cell that develops one or more open ends from which a basipetal succession of conidia develops without an increase in length of the phialide itself.
Physiogenic Disease Caused by unfavorable environmental factors.
Physiologic Races Pathogens of same variety and species structurally the same but differing in physiological behavior, especially in ability to parasitize a given host.
Phytopathology Plant pathology, science of plant disease.
Phytoplasma A MLO that is pathogenic to plants.
Pileus Hymenium-supporting part of a fruit body of a higher fungus; the cap of a mushroom.
Primary Infection First infection by a pathogen after going through a resting or dormant period.
Prokaryotic Organisms which lack a true nucleus; includes bacteria and mycoplasmalike organisms.
Promycelium Basidium of rusts and smuts.
Pulvinate Cushionlike in form.
Pycnidium, pl. Pycnidia Flasklike fruiting body containing conidia.
Pycnium Spermagonium in the rusts, the 0 stage, resembling a pycnidium.

Resistance Ability of a host plant to suppress or retard activity of a pathogen.
Resting Spore A spore, often thick-walled, that can remain alive in a dormant condition for some time, later germinating and capable of initiating infection.
Resupinate Flat on the substratum with hymenium on outer side.
Rhizoid Rootlike structure.
Rhizomorph A cordlike strand of fungus hyphae.

Ring Spot Disease symptoms characterized by yellowish or necrotic rings with green tissue inside the ring, as in virus diseases.
Roguing Removal of undesired individual plants.
Rosette Disease symptom with stems shortened to produce a bunchy growth habit.
Russet Brownish roughened areas on skins of fruit, from abnormal production of cork caused by disease, insect, or spray injury.
Rust A fungus, one of the Uredinales, causing a disease also known as rust.

Saprophyte An organism that feeds on lifeless organic matter.
Scab Crustlike disease lesion; or a disease in which scabs are prominent symptoms.
Sclerotium, pl. Sclerotia Resting mass of fungus tissue, often more or less spherical, normally having no spores in or on it.
Scorch Burning of tissue, from infection or weather conditions.
Scutellum Plate or shieldlike cover, as in Microthyriales.
Septate Having cross-walls, septa.
Sessile Having no stem.
Seta, pl. Setae A stiff hair, or bristle, generally dark-colored.
Shot Hole A disease symptom in which small round fragments drop out of leaves, making them look as if riddled by shot.
Sign Any indication of disease other than reaction of the host plant – pores, mycelium, exudate, or fruiting bodies of the pathogen.
Slurry Thick suspension of chemical; used for seed treatment.
Smut A fungus of the Ustilaginales, characterized by sooty spore masses; the name also used for the disease caused by the smut.
Sooty Mold Dark fungus growing in insect honeydew.
Sorus, pl. Sori Fungus spore mass, especially of rusts and smuts; occasionally, a group of fruiting bodies.
Species One sort of plant or animal; abbreviated as "sp." singular, and "spp." plural. A genus name followed by sp. means that the particular species is undetermined. Spp. following a genus name means that several species are grouped together without being named individually.
Spermagonium Walled structure in which spermatia are produced, a pycnium.
Spermatium, pl. Spermatia A sex cell (+ or −), a pycniospore.
Sporangiole Small sporangium without a columella and with a small number of spores.
Sporangiophore Hypha bearing a sporangium.
Sporangium Organ producing nonsexual spores in a more or less spherical wall.
Spore A single-to many-celled reproductive body, in the fungi and lower plants, which can develop a new plant.
Sporidium, pl. Sporidia Basidiospore of rusts and smuts.
Sporodochium, pl. Sporodochia Cluster of conidiophores interwoven on a stroma or mass of hyphae.
Sporophore Spore-producing or supporting structure – fruit body; used especially in the Basidiomycetes.
Sporulate To produce spores.

Sprayer Apparatus for applying chemicals in liquid form.
Sterigma, pl. Sterigmata Projection for supporting a spore.
Stipe A stalk. *Stipitate.* Stalked.
Strain An organism or group of organisms differing in origin or minor aspects from other organisms of same species or variety.
Stroma, pl. Stromata Mass of fungus hyphae often including host tissue containing or bearing spores.
Subiculum, Subicle Netlike woolly or crustlike growth of mycelium under fruit bodies.
Substrate The substance or object on which a saprophytic organism lives and from which it gets nourishment.
Suscept A living organism attacked by, or susceptible to, a given disease or pathogen; in many cases a more precise term than host but less familiar.
Susceptible Unresistant, permitting the attack of a pathogen.
Swarmspore Zoospore.
Synnema, pl. Synnemata Groups of hyphae sometimes joined together, generally upright and producing spores; coremium.
Systemic Term applied to disease in which single infection leads to general spread of the pathogen throughout the plant body; or to a chemical that acts through the vascular system.

Teleomorph State of life-cycle in which spores are formed after nuclear fission.
Teliospore Winter or resting form of rust spore, from which basidium is produced.
Telium Sorus producing teliospores.
Thallophyte One of the simpler plants, belonging to the algae, bacteria, fungi, slime molds, or lichens.
Thallus Vegetative body of a thallophyte.
Tolerant Capable of sustaining disease without serious injury or crop loss.
Toxin Poison formed by an organism.
Tylosi, pl. Tyloses Cell outgrowth into cavity of xylem vessel, plugging it.

Urediospore Summer spore of rusts; one-celled, verrucose.
Uredium Sorus producing urediospores.

Valsoid Having groups of perithecia with beaks pointing inward, or even parallel with surface, as in valsa.
Vector An agent, insect, man, etc., transmitting disease.
Vein-banding Symptom of virus disease in which regions along veins are darker green than the tissue between veins.
Verrucose With small rounded processes or warts.
Viroid The smallest known viruslike infectious agent having no protein coat and only a small amount of nucleic acid.
Virulent Highly pathogenic; with strong capacity for causing disease.
Viruliferous Virus-carrying; term applied particularly to virus-laden insects.
Virus An obligate parasite capable of multiplying in certain hosts, ultramicroscopic, recognizable by the effects produced in infected hosts. Has nucleic acid with protein coat.

Wilt Loss of freshness or drooping of plants due to inadequate water supply or excessive transpiration; a vascular disease interfering with utilization of water.

Witches' Broom Disease symptom with abnormal brushlike development of many weak shoots.

Yellows Term applied to disease in which yellowing or chlorosis is a principal symptom.

Zoospore A swimming spore, swarmspore, capable of independent movement.

Zygomycetes Subclass of the Phycomycetes, characterized by gametes of equal size.

Zygospore Resting spore formed from the union of similar gametes.

MIX
Papier aus verantwortungsvollen Quellen
Paper from responsible sources
FSC® C105338

If you have any concerns about our products,
you can contact us on
ProductSafety@springernature.com

In case Publisher is established outside the EU,
the EU authorized representative is:
Springer Nature Customer Service Center GmbH
Europaplatz 3, 69115 Heidelberg, Germany

Printed by Libri Plureos GmbH
in Hamburg, Germany